AF368955

Data Processing and Modeling on Volcanic and Seismic Areas

Data Processing and Modeling on Volcanic and Seismic Areas

Editors

Alessandro Bonforte
Flavio Cannavò

MDPI • Basel • Beijing • Wuhan • Barcelona • Belgrade • Manchester • Tokyo • Cluj • Tianjin

Editors
Alessandro Bonforte
Istituto Nazionale di
Geofisica e Vulcanologia,
Sezione di Catania—
Osservatorio Etneo
Italy

Flavio Cannavò
Istituto Nazionale di Geofisica e Vulcanologia (INGV),
Osservatorio Etneo
Italy

Editorial Office
MDPI
St. Alban-Anlage 66
4052 Basel, Switzerland

This is a reprint of articles from the Special Issue published online in the open access journal *Applied Sciences* (ISSN 2076-3417) (available at: https://www.mdpi.com/journal/applsci/special_ issues/volcanic_seismic).

For citation purposes, cite each article independently as indicated on the article page online and as indicated below:

LastName, A.A.; LastName, B.B.; LastName, C.C. Article Title. *Journal Name* **Year**, *Volume Number*, Page Range.

ISBN 978-3-0365-3853-2 (Hbk)
ISBN 978-3-0365-3854-9 (PDF)

Cover image courtesy of Alessandro Bonforte and Flavio Cannavò.

Contents

About the Editors

Alessandro Bonforte working at the Istituto Nazionale di Geofisica e Vulcanologia since 2003. Mainly focused on multi-parametric ground deformation studies using satellite and ground-based techniques on active volcanic and tectonic areas and multidisciplinary data integration. He is responsible for the GNSS mobile and periodic networks at Osservatorio Etneo, member of the IngvVulcani national group for communication and divulgation on blog and media, and chair of the IAG/IAVCEI Joint Commission for Volcano Geodesy.

Flavio Cannavò has been working at INGV Etnean Observatory since 2007, conducting his research in different fields. His areas of expertise include geophysical and ground deformation data analysis and modelling, machine learning and data science, which are mainly applied to volcanology.

Preface to "Data Processing and Modeling on Volcanic and Seismic Areas"

The Earth Sciences are continuously developing from a descriptive "naturalistic" science towards an increasingly physical and "numerical" science. Quantitative data are being collected from a huge variety of disciplines in a dramatically increasing way, from a huge variety of sensors on Earth and remote satellites. This growth requires a parallel progression of the systems to archive, manage, process and interpret data. A large scientific and technological effort is devoted to extracting as much information as possible from the collected data.

The Special Issue collects up-to-date research on the processing and modeling of Earth Sciences data, and addresses some the broad, currently challenging problems in volcanic and seismic topics.

The collection reveals the great need to manage the growing complexity of data and the treatment of data, to try to describe the natural phenomena that are considered here.

The issue is aimed at any researcher who is curious about the state of the art in the use of data to analyze and model processes related to volcanoes or earthquakes.

Alessandro Bonforte and Flavio Cannavò
Editors

Editorial

Special Issue "Data Processing and Modeling on Volcanic and Seismic Areas"

Alessandro Bonforte * and Flavio Cannavò

Istituto Nazionale di Geofisica e Vulcanologia, Sezione di Catania–Osservatorio Etneo, Piazza Roma, 2, 95125 Catania, Italy; flavio.cannavo@ingv.it
* Correspondence: alessandro.bonforte@ingv.it

Citation: Bonforte, A.; Cannavò, F. Special Issue "Data Processing and Modeling on Volcanic and Seismic Areas". *Appl. Sci.* **2021**, *11*, 10759. https://doi.org/10.3390/app1122 10759

Received: 14 October 2021
Accepted: 8 November 2021
Published: 15 November 2021

Publisher's Note: MDPI stays neutral with regard to jurisdictional claims in published maps and institutional affiliations.

Volcanology, seismology and Earth Sciences in general, like all quantitative sciences, are increasingly dependent on the quantity and quality of data acquired. In recent decades, a marked evolution has characterized Earth sciences towards a greater use of analytical and numerical approaches, shifting these fields from the natural to the physical sciences.

Understanding the physical behavior of active volcanoes and faults is critical to assess the hazards affecting the population living close to active volcano and seismic areas, and thus to mitigate the risks posed by those threats [1,2]. The knowledge of a physical process requires the acquisition of a huge amount of information (data) on that particular phenomenon.

Today, different kinds of data record the processes that operate in volcanic and tectonic systems and provide insights that can lead to improved predictions of potential hazards, both immediate and long term. The geoscience community has collected an enormous wealth of data that require further analysis. The diversity and quantity of these geoscience data and collections continue to expand [3].

The increasing amount of data and the availability of new technologies and instrumentation at an ever-greater rate open new frontiers and challenges for acquiring, transmitting, archiving, processing and analyzing the newly available datasets. Guo [4] predicted growth for the general digital universe size of factor 10 from 2016 to 2025. Among all digital data, scientific data are those relevant to the observation of natural phenomena and characterized by non-repeatability, high uncertainty, high dimensionality and a high degree of computational complexity [4]. This means that scientific data need to be well preserved, due to the non-repeatability, and implies a parallel growth of processing capabilities to be well exploited. Cheng et al. [5] highlighted the striking growth of Earth Science data from molecular to astronomical scales and the increasing use of supercomputing tools for supporting geoscience research. The authors evidence how, with the continuously increasing availability of digital data, Earth Sciences are also turning from the traditional question-driven or problem-driven approach, where scientists seek to find answers to known questions, to the new data-driven one where scientists apply a data discovery process that might find answers to still unknown questions.

In agreement with Cheng et al. [5], we believe that new integrated multi-disciplinary knowledge systems and new data discovery techniques for handling and mining big data for knowledge discovery would spur the integration of transdisciplinary and multi-dimensional Earth science data. Furthermore, this will help the transition from a narrow focus on separate disciplines to a holistic, comprehensive and integrative focus of the different disciplines linked to the Earth Sciences.

With this aim, for this special issue titled "Data Processing and Modeling on Volcanic and Seismic Areas", we invited articles on all aspects of solid Earth Science that made use of data to analyze and model processes related to volcanoes or earthquakes.

Manuscripts with various types of analyses, including volcanic ground deformation modeling, seismic swarm characterization and volcanic gas measurement, have been proposed and published.

The collection provides an insight into the enormous need for increasingly complex data analysis and modeling techniques to try to describe the natural phenomena here considered.

This special issue was introduced to collect the latest research on the processing and modeling of Earth Sciences data, and to address challenging problems with all topics related to volcanoes and seismic areas. Various subjects have been addressed in this collection, mainly on data processing for volcanic studies (three papers), tectonics (two papers) and one paper on data analysis of a new instrument to measure gases.

The first contribution to this collection [6] reports the results of the processing and combination of high-rate and low-rate geodetic data for revealing the dynamics underlying violent volcanic eruptions at Mount Etna. This study evidences the wide spectrum of ground deformation produced by these phenomena, to be investigated, processed and modeled in order to generate a picture of the feeding system of the volcano and better understand its dynamics and rates of magma transfer in the upper crust.

Another contribution focuses on volcanoes [7]: the authors exploit 20 years of high temporal resolution satellite Thermal Infra-Red (TIR) data collected over three active volcanoes (Etna, Shishaldin and Shinmoedake). They present the results of an analysis of this dataset performed through a preliminary RST (Robust Satellite Techniques) algorithm implementation to TIR data from the Advanced Spaceborne Thermal Emission and Reflection Radiometer (ASTER). This approach ensures efficient identification and mapping of volcanic thermal features even of a low intensity level, which is also useful in the perspective of an operational multi-satellite observing system.

The contribution by Woohyun Son et al. [8] proposes specific depth-domain data processing of migration velocity analysis (MVA) of seismic data collected during a survey on a saline aquifer sediment in the Southern Continental Shelf of Korea. This analysis allowed the authors to identify and determine the precise depth of a basalt flow that could act as a cap rock for CO_2 storage beneath the aquifer. The investigation, through the geological model obtained from both time- and depth-domain processing, provides suitable information for locating the best drilling sites for CO_2 injection, maximizing the storage volume.

In volcanic areas, gases represent important physical evidence of volcanic processes that need to be measured. Parracino et al. [9] have shown how novel range-resolved DIAL-Lidar (Differential Absorption Light Detection and Ranging) could herald a new era in the observation of long-term volcanic CO_2 gases.

An accurate and integrated analysis of different types of data such as GNSS, seismic and MT-InSAR, has led, in the work by Gatsios et al. [10], to a first account of deformation processes and their temporal evolution over recent years for Methana (Greece), thus providing initial information to feed into a volcano baseline hazard assessment and monitoring system.

Seismic data are among the most important data to understand the dynamics of the Earth's interior. A consistent analysis of a seismic swarm allowed Kostoglou et al. [11] to shed more light on the regional geodynamics of the Kefalonia Transform Fault Zone (Greece), and to follow the temporal evolution of the b-value to distinguish between foreshock and aftershock behaviors.

Author Contributions: A.B. and F.C. contributed to the design and implementation of the special issue and to the writing of the manuscript. All authors have read and agreed to the published version of the manuscript.

Funding: This research received no external funding.

Institutional Review Board Statement: Not applicable.

Informed Consent Statement: Not applicable.

Conflicts of Interest: The authors declare no conflict of interest.

References

1. Segall, P. Volcano deformation and eruption forecasting. *Geol. Soc. Lond. Spec. Publ.* **2013**, *380*, 85–106. [CrossRef]
2. Wu, Z.; Zhang, Y.; Goebel, T.H.; Huang, Q.; Williams, C.A.; Xing, H.; Rundle, J.B. Continental Earthquakes: Physics, Simulation, and Data Science—Introduction. *Pure Appl. Geophys.* **2020**, *177*, 1–8. [CrossRef]
3. Fagents, S.A.; Gregg, T.K.; Lopes, R.M. (Eds.) *Modeling Volcanic Processes: The Physics and Mathematics of Volcanism*; Cambridge University Press: Cambridge, UK, 2013.
4. Guo, H. Big Earth data: A new frontier in Earth and information sciences. *Big Earth Data* **2017**, *1*, 4–20. [CrossRef]
5. Cheng, Q.; Oberhänsli, R.; Zhao, M. A new international initiative for facilitating data-driven Earth science transformation. *Geol. Soc. Lond. Spec. Publ.* **2020**, *499*, 225–240. [CrossRef]
6. Bonforte, A.; Cannavò, F.; Gambino, S.; Guglielmino, F. Combining High-and Low-Rate Geodetic Data Analysis for Unveiling Rapid Magma Transfer Feeding a Sequence of Violent Summit Paroxysms at Etna in Late 2015. *Appl. Sci.* **2021**, *11*, 4630. [CrossRef]
7. Genzano, N.; Marchese, F.; Neri, M.; Pergola, N.; Tramutoli, V. Implementation of Robust Satellite Techniques for Volcanoes on ASTER Data under the Google Earth Engine Platform. *Appl. Sci.* **2021**, *11*, 4201. [CrossRef]
8. Son, W.; Cheong, S.; Lee, C.; Kang, M. Imaging Top of Volcanic Mounds Using Seismic Time-and Depth-Domain Data Processing. *Appl. Sci.* **2021**, *11*, 4244. [CrossRef]
9. Parracino, S.; Santoro, S.; Fiorani, L.; Nuvoli, M.; Maio, G.; Aiuppa, A. The BrIdge voLcanic LIdar-BILLI: A Review of Data Collection and Processing Techniques in the Italian Most Hazardous Volcanic Areas. *Appl. Sci.* **2020**, *10*, 6402. [CrossRef]
10. Gatsios, T.; Cigna, F.; Tapete, D.; Sakkas, V.; Pavlou, K.; Parcharidis, I. Copernicus sentinel-1 MT-InSAR, GNSS and seismic monitoring of deformation patterns and trends at the Methana Volcano, Greece. *Appl. Sci.* **2020**, *10*, 6445. [CrossRef]
11. Kostoglou, A.; Karakostas, V.; Bountzis, P.; Papadimitriou, E. The February–March 2019 Seismic Swarm Offshore North Lefkada Island, Greece: Microseismicity Analysis and Geodynamic Implications. *Appl. Sci.* **2020**, *10*, 4491. [CrossRef]

applied sciences

Article

Combining High- and Low-Rate Geodetic Data Analysis for Unveiling Rapid Magma Transfer Feeding a Sequence of Violent Summit Paroxysms at Etna in Late 2015

Alessandro Bonforte *, Flavio Cannavò, Salvatore Gambino and Francesco Guglielmino

Istituto Nazionale di Geofisica e Vulcanologia, Sezione di Catania-Osservatorio Etneo, Piazza Roma, 2-95125 Catania, Italy; flavio.cannavo@ingv.it (F.C.); salvatore.gambino@ingv.it (S.G.); francesco.guglielmino@ingv.it (F.G.)
* Correspondence: alessandro.bonforte@ingv.it; Tel.: +39-095-716-5800

Abstract: We propose a multi-temporal-scale analysis of ground deformation data using both high-rate tilt and GNSS measurements and the DInSAR and daily GNSS solutions in order to investigate a sequence of four paroxysmal episodes of the Voragine crater occurring in December 2015 at Mt. Etna (Italy). The analysis aimed at inferring the magma sources feeding a sequence of very violent eruptions, in order to understand the dynamics and to image the shallow feeding system of the volcano that enabled such a rapid magma accumulation and discharge. The high-rate data allowed us to constrain the sources responsible for the fast and violent dynamics of each paroxysm, while the cumulated deformation measured by DInSAR and daily GNSS solutions, over a period of 12 days encompassing the entire eruptive sequence, also showed the deeper part of the source involved in the considered period, where magma was stored. We defined the dynamics and rates of the magma transfer, with a middle-depth storage of gas-rich magma that charges, more or less continuously, a shallower level where magma stops temporarily, accumulating pressure due to the gas exsolution. This machine-gun-like mechanism could represent a general conceptual model for similar events at Etna and at all volcanoes.

Keywords: volcano geodesy; multidisciplinary monitoring; paroxysms; lava fountain; volcanic eruption; modeling; tilt; GPS; InSAR; ground deformation

Citation: Bonforte, A.; Cannavò, F.; Gambino, S.; Guglielmino, F. Combining High- and Low-Rate Geodetic Data Analysis for Unveiling Rapid Magma Transfer Feeding a Sequence of Violent Summit Paroxysms at Etna in Late 2015. *Appl. Sci.* **2021**, *11*, 4630. https://doi.org/10.3390/app11104630

Academic Editor: Stuart Hardy

Received: 19 March 2021
Accepted: 10 May 2021
Published: 19 May 2021

Publisher's Note: MDPI stays neutral with regard to jurisdictional claims in published maps and institutional affiliations.

1. Introduction

Mt. Etna is the largest and highest volcano in continental Europe. It is located in the central Mediterranean Sea, in eastern Sicily (Italy), just north of the city of Catania (Figure 1). It lies on continental crust, over the collisional belt between the Nubia and the Eurasia plates. The northern side of Etna lies on the external units of the Apennine-Maghrebian chain, while the southern foot lies on the foredeep deposits, where the foreland bends northwards beneath the chain. This geodynamic framework is further complicated by the presence of the Maltese Escarpment, on the eastern foot of the volcano, separating the continental crust in the west from the oceanic one of the Ionian basin in the east. Just east of Etna, in fact, the collisional dynamics are no longer characterized by continental crust collision but by the subduction of the oceanic Ionian beneath the Calabrian Arc.

The Mt. Etna volcano is also one of the most active volcanoes on our planet. It erupts almost continuously, usually from its four summit craters (Figure 2), named Voragine (VOR), Northeast Crater (NEC), Bocca Nuova (BN) and Southeast Crater (SEC), where, since 2011, a new cone has formed and rapidly grown on its eastern slope, named the New Southeastern Crater (NSEC) [1]. Less frequently, magma rises up from lateral fissures and eruptions occur at lower altitudes on the volcano's flanks, producing more hazardous lava flows for those villages and rural areas surrounding the volcano.

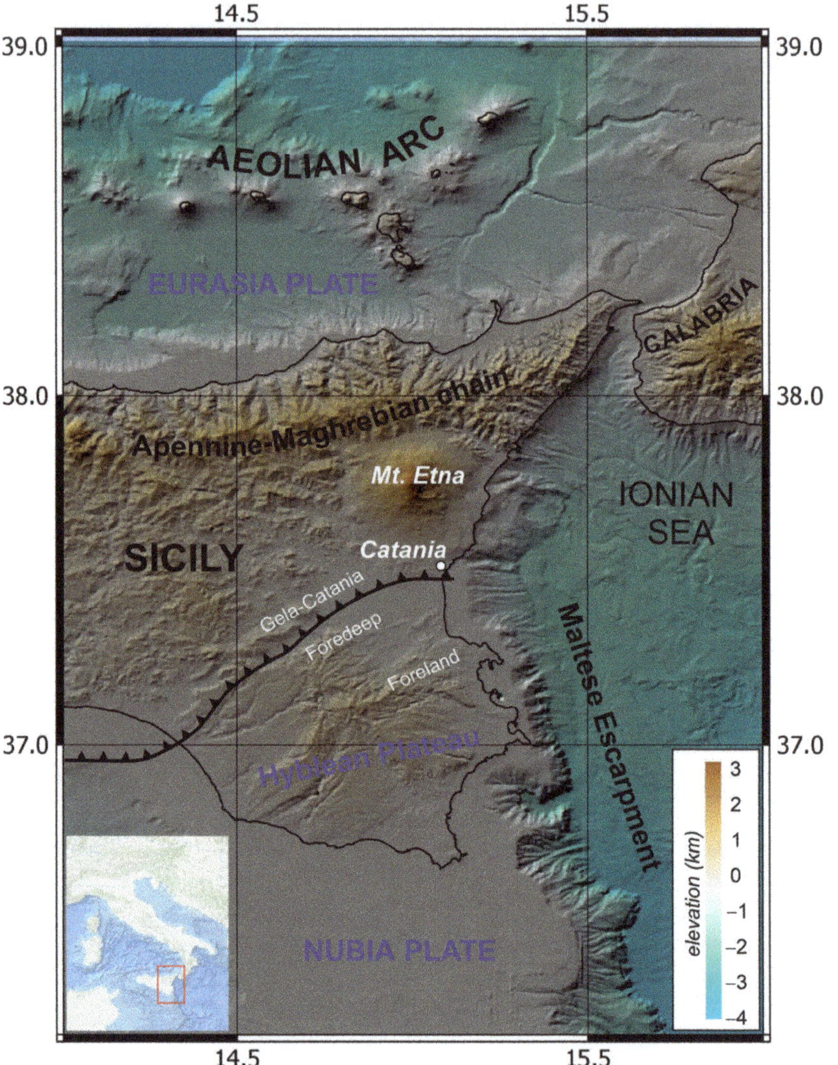

Figure 1. Location of Mount Etna in the central Mediterranean in the inset; in the main figure, the detail of the geodynamic framework x surrounding the volcano (zoom on eastern Sicily–Calabrian arc, red box in the inset). The shaded-relief basemap is a combination of Emodnet bathymetry (http: //www.emodnet-bathymetry.eu, accessed on 16 March 2021) and SRTM 30 plus topography (Shuttle Radar Tomography Mission, http://topex.ucsd.edu/WWW_html/srtm30_plus.html, accessed on 16 March 2021).

Eruptive activity at summit craters is usually characterized by strombolian activity of varying intensity, up to the more energetic fire fountains and high-energy paroxysms.

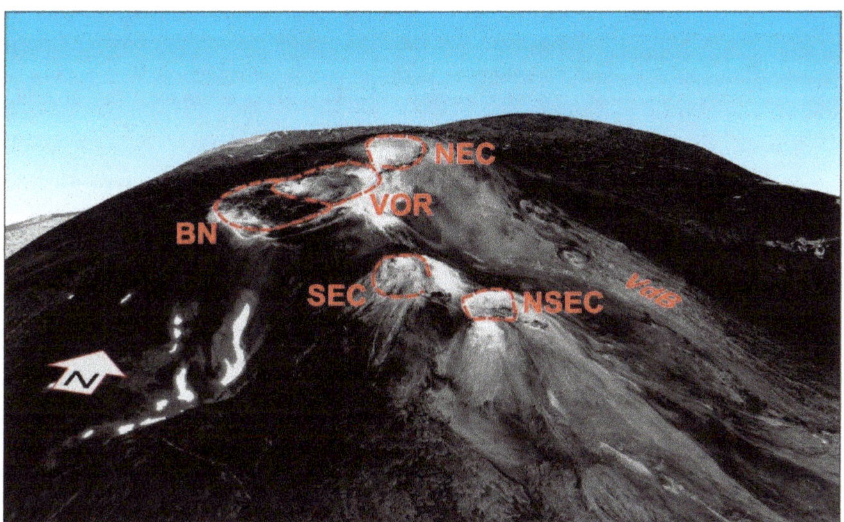

Figure 2. Mt. Etna summit craters: Voragine (VOR), Northeast Crater (NEC), Bocca Nuova (BN), Southeast Crater (SEC) and New Southeastern Crater (NSEC). The 3D surface map is a combination of very-high resolution optical images acquired by Pleiades satellites on 20160603 and an SRTM DEM.

2. The December 2015 Activity

Before the 2015 events, eruptive activity at Etna consisted of a series of paroxysmal episodes at NSEC, which produced 50 lava fountains from 2011 to 2014. Conversely, after several years of mainly degassing activity, starting on 3 December 2015, the VOR crater underwent an extraordinary sequence of four very energetic paroxysms until December 5. Such a rapid sequence of such powerful events was never observed before on Etna, especially if considering that, in the short period of time between 2 and 18 December 2015, all the four summit craters produced explosive activity. The most powerful event was the first one, while the following three events showed a decreasing energy. All showed the same pattern starting from strombolian activity and rapidly evolving to lava fountains in a few minutes, producing very high lava fountains, launching incandescent clasts to more than 3 km above the crater rim and producing an eruptive column reaching 14 km a.s.l., then decreasing again to strombolian activity before ending [2]. In detail, on 3 December, at about 02:00 UTC the explosive activity at the Voragine crater was characterized by a very fast increase in the initial strombolian activity, which at about 2:40 UTC culminated in a lava fountain (I episode) lasting 90 min, reaching a maximum height of the lava jet thrust of 4100 m above the vent [3], and then gradually passed again to strombolian activity. The same sequence was recorded twice on December 4; the lava fountains occurred at 9:15 UTC (II episode ca. 80 min) and at 20:30 UTC (III episode with a duration of 90 min). Finally, in the early afternoon of December 5 from 13:30 UTC, a slower increase in explosive activity was observed at the VOR crater, switching to fountain activity at about 15:00 UTC for ca. one hour and a half (IV episode). The entire sequence at VOR ended with the last episode in the afternoon of December 5, lasting only 38 h, followed by weak strombolian activity and short lava flows from NSEC until December 8. Authors in [2] estimated a volume for pyroclastic deposits of about 7.1×10^6 m^3, resulting in an average of about 1.8×10^6 m^3 erupted during each paroxysm.

Such a violent eruptive phase, characterized by a sequence of paroxysms ranging from lava fountains to sub-plinian explosions, are not uncommon. In the first half of 2000, Mt. Etna generated 66 paroxysmal episodes [4] from the SEC, while the VOR crater produced less but very powerful activity, a sub-plinian eruption in July 1998 [5] and another very strong paroxysm one year later, in September 1999 [6]. Between 2011 and 2015, Mt. Etna saw more than 50 paroxysmal episodes from its summit vents. Furthermore, these kinds of eruptions have also been observed on other basaltic volcanoes, from the first observations and studies of Pelèean eruptions at Manam (Papua New Guinea) in the 1960s to the more recent sequences at Kilawea (Hawaii) in the 1980s or at Fuego (Guatemala) in 2018.

3. Materials and Methods

To perform our analysis of the source feeding such a violent crisis on Etna, we exploited the multi-parametric dataset coming from the permanent ground deformation networks on the volcano (Figure 3) and the DInSAR images pair acquired across the entire eruptive period by the ESA Sentinel-1 satellite.

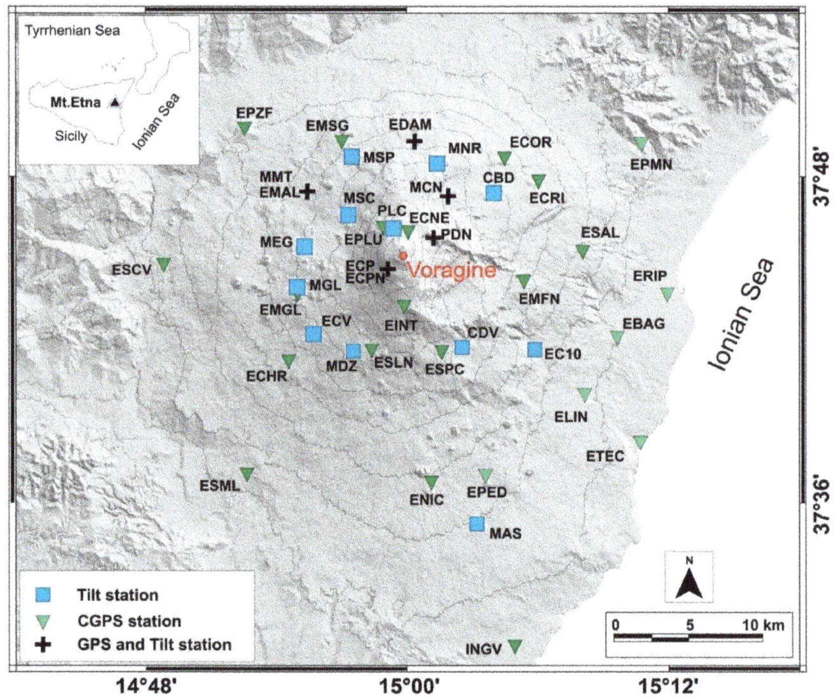

Figure 3. Map of the permanent tilt and GNSS networks operating on Mt. Etna.

3.1. Tilt Data

On Mt. Etna, surface local inclination variations are continuously monitored by using instruments able to detect tilt changes with high-precision data.

In December 2015, the Mt. Etna permanent tilt network consisted of 14 borehole instruments and a long-base device (PDN, Figure 3). Over the last ten years, INGV improved the network through the installation of sensors at greater depth: currently, eight stations are at 27–30 m depth and six are 10 m deep, including three summit stations.

The deeper stations use biaxial self-leveling instruments with very high-precision (10^{-8}–10^{-9} radians) electrolytic bubble sensors for measuring the angular movement and magnetic compasses that are able to detect tides. Almost all stations have a sampling rate of 1 min. A description of the tilt instrumental characteristics is provided by [7–9].

Continuous tiltmeters recorded variations during the four fountain episodes occurring on 3–5 December 2015. Tilt changes during the first, most powerful lava fountain were recorded at all the stations showing values comprising a few tenths to a couple of microradians with a mean value of 1.1 (Figure 4). The magnitude of episode II was slightly lower (mean value of 0.85 microrads), while the successive episodes had a significantly smaller amplitude (mean values, respectively, of 0.5 and 0.35 microradians).

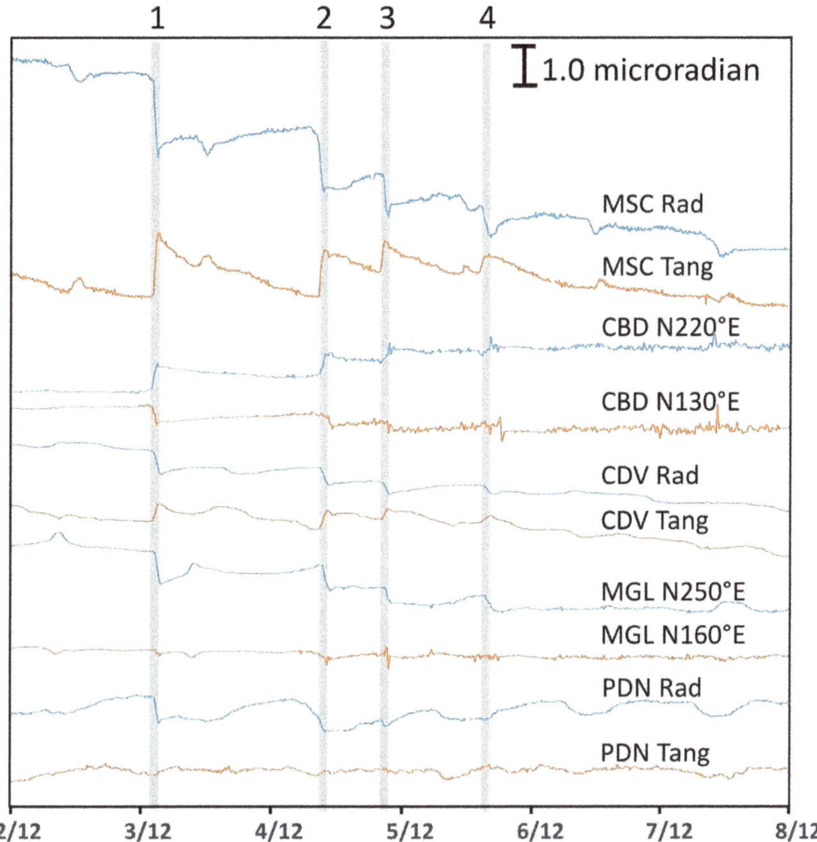

Figure 4. Tilt components recorded at several stations during the fountain episodes. Grey lines highlight the 4 paroxysms evidencing associated changes. Radial (Rad) components are directed toward summit craters and positive variations indicate an upward tilt. Oriented components (e.g., CBD N220°E) indicate a downward tilt along the indicated direction.

The first change occurred on December 3 at about 02:30 UTC and all the four episodes showed a mean duration of about 1 h. The pattern of the ground tilt was the same for each event, even if its magnitude decreased from the first to last episode, evidencing similar sources (Figure 4). At the end of the entire sequence, the tilt showed the same trends as before the fountains.

The tilt vectors showed a general lowering toward an area positioned just west of the Mt. Etna summit (Figure 5), suggesting a repeated occurrence of deflation processes.

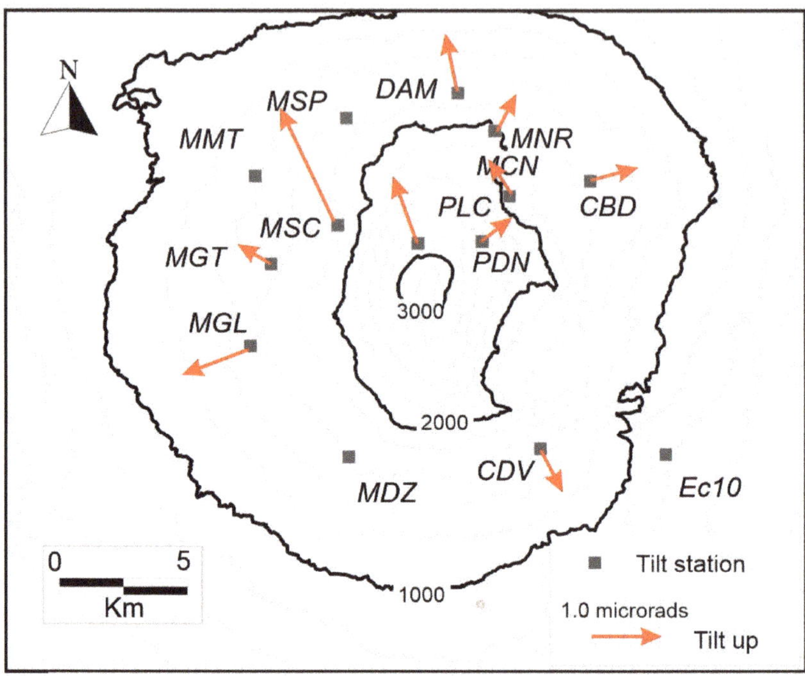

Figure 5. Observed tilt vectors cumulated during the first lava fountain indicating an edifice deflation. MMT, MDZ, MSP and EC10 stations were out of order.

3.2. GPS Data

Geodetic monitoring on Etna started early in the 1970s with pioneristic trilateration networks; since then, a continuous evolution of methods and technologies has led to the current configuration and complexity designed for several applications and studies at different space and time scales [10–12]. Now, with its 39 permanent stations, the Etnean GPS network is one of the largest deployed on an active volcano (Etn@net, [13]). For surveillance purposes, most of the stations are acquired every second and processed in real-time at high-rate (1 Hz) by using the epoch-by-epoch algorithm of Geodetics ® RTD software package [14].

High-rate GPS time series contain both systematic errors (e.g., multipath) and unmodeled stochastic noise, which can be in the same order of magnitude of the displacement signal. In this study, we applied an ad-hoc strategy in order to minimize the noise level of the time series. To this end, the high-rate data from continuous GPS stations belonging to the INGV Etnean GPS network were processed in kinematic mode (which means assuming that the station is continuously moving) by using three different algorithms: (i) precise epoch-by-epoch solutions [14], by using the Geodetics RTD software which provides independent estimates of positions for each epoch of dual-frequency; (ii) Precise Point Positioning (PPP) solutions [15] by using the GIPSY software; and (iii) differential phase kinematic [16] by using the TRACK software [17], which uses the ionosphere-free LC observable double-differences with respect to nearby fixed stations. All the available raw GPS data were decimated at 30 s and the solutions were sub-sampled with medians in windows of 10 min. As a reference fixed station we chose ENIC (see Figure 3).

To highlight the displacement signals in the noisy time series of GPS components, we calculated the baseline variations; this allowed us to significantly reduce common sources of noise (e.g., atmosphere, constellation geometry, etc.) which were predominant in the high-rate GPS data. In order to enhance the useful variations against the background noise and to further reduce the geodetic correlated noise, we combined the obtained time series from the different processing software by a weighted mean optimized to minimize the resulting associated uncertainties [18]. This can be obtained by considering the inverses of estimated variances as weights (covariance matrix is also provided in the Supplementary Materials).

Figure 6 shows the time series calculated as the mean of all the considered baseline relative variations, expressed in parts-per-million (ppm). To estimate the displacements and their associated uncertainties, a piecewise linear regression was carried out on the baseline time series. The choice of using a piecewise line to infer the slopes followed a physical time continuity. The break points for the segmented line were set to the beginning and ends of the paroxysms. A robust regression algorithm [19] was applied, by which the piecewise regression line was made to fit the data set as closely as possible while minimizing the sum of squares of the differences.

To verify the stability of the regression results, we tried to linearly fit independently the episodes (e.g., yellow line in Figure 6C). It is worth noting that the variations are quite similar and comparable with the ones obtained by piecewise regression. The only discrepancy with our approach is the period in between the second and third paroxysms when the independent linear fit does not show the same positive step as in the piecewise line. However, this does not affect the coherent deflective behavior during the third paroxysm.

To infer the sources, we considered the total variations of the regression lines in the eruptive periods.

Figure 6 shows that the eruptive periods were characterized by a shrinking, albeit small, of the baselines crossing the summit area. This indicates a contraction of the volcano edifice. The deformation strength was higher for the first event than for the last one, indicating a progressive decay of the eruption energy. In some cases, it is possible to also see a small inflation phase prior to the eruptions (a lengthening of some baselines crossing the summit area).

It is worth noting that such variations are within the error bars of the data; nevertheless, they are coherent among the baselines and follow the expected volcanic behavior, thus can be considered reliable for a source estimation.

3.3. DInSAR Data

Sentinel 1A SAR (Synthetic Aperture Radar) images were acquired in TopSAR (Terrain Observation with Progressive Scans SAR) Interferometric Wide mode, during ascending orbit over Mt. Etna on December 2 and successively on the next ascending pass on December 14, after the four paroxysms occurred. The differential interferogram (DInSAR) combining the two subsequent passes, encompassing the entire eruptive activity at the central crater, was processed with the GAMMA software, and in order to remove the topographic phase, a Shuttle Radar Topography Mission (SRTM) Version 4 digital elevation model with 3 arc-sec resolution was used, and is shown in Figure 7. This interferogram is very useful to draw the total cumulated ground deformation pattern affecting the volcano after such a violent eruptive phase. The interferogram clearly shows that, from December 2 to 14, the entire upper part of the volcano moved away from the satellite, meaning that it moved downwards and/or eastwards (the satellite views from the west with an incidence angle of about 38° from the vertical). The downwards and eastwards movement of the ground is more evident on the upper western side of the volcano, while on the eastern side, the downwards movements could be partially compensated by a westward one, coherently with a deflation/contraction pattern.

Another small area showing local deformation is detected by DInSAR on the middle NE flank of the volcano, related to the uppermost part of the well-known Provenzana-Pernicana fault system [20,21] that was affected by a seismic swarm on 8 December 2015, following the abrupt deflation produced by the four eruptions. Additionally, in this area, the ground deformation pattern, moving away from the satellite on the southern side of the fault (hangingwall), reveals a downwards/eastwards ground motion, in agreement with the usual oblique normal and left-lateral displacement of this fault system, clearly related to the flank dynamics of the volcano [22–25].

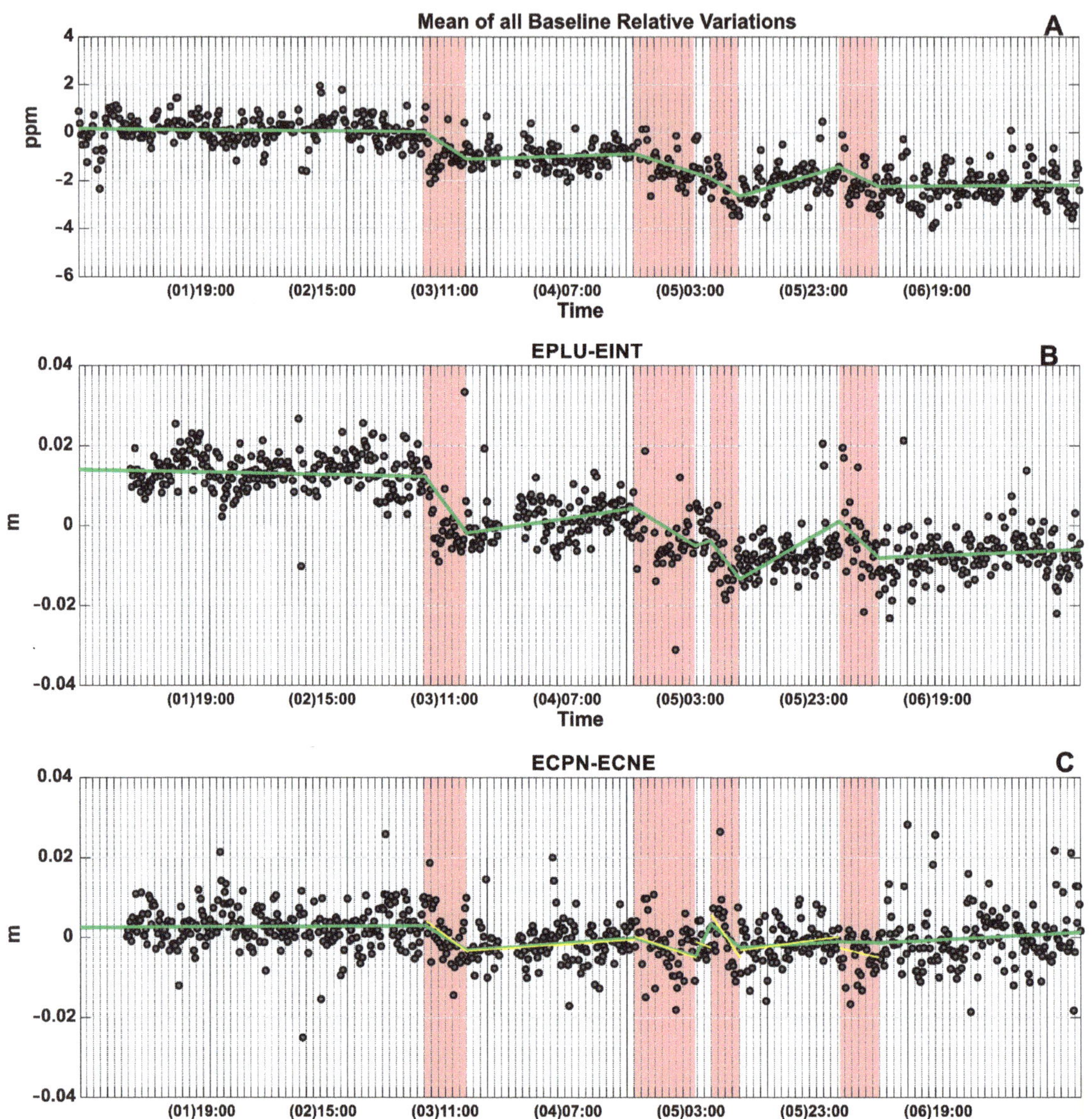

Figure 6. Time series of GPS baselines. Black dots are the raw baseline measurements. Red stripes represent the eruptive periods. Green lines are the optimal piecewise linear interpolation of time series for the considered periods. (**A**) Average series of the relative variations (in ppm) of all the considered baselines, that can be read as the measured mean micro-strain. (**B**) Time series of baseline variation between EPLU and EINT GPS stations. (**C**) Time series of baseline variation between ECNE and ECPN GPS stations; the yellow segments represent the linear fit for each independent period.

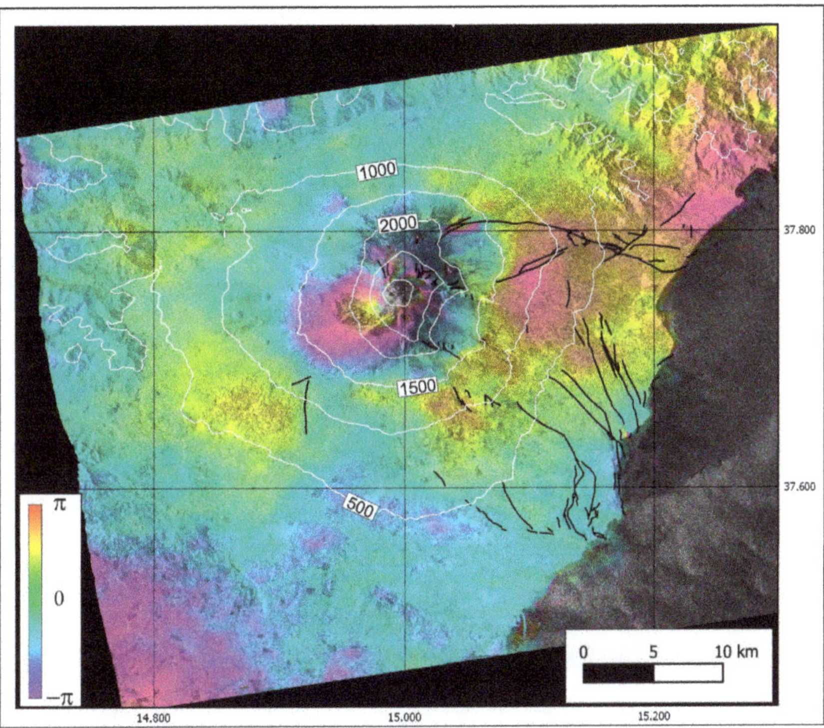

Figure 7. DInSAR interferogram of the ascending orbit pair from 2 to 14 December 2015. A fringe (28 mm) of deflation is visible, affecting the whole upper part of the volcano, above 1500–2000 m a.s.l., and across the upper Pernicana-Provenzana fault system on the middle NE flank. The topography is represented with white contour lines (every 500 m).

3.4. SISTEM—DInSAR and GPS Integrated Data

To estimate the total deformation occurring at the volcano during the entire crisis, we produced a high-resolution map by integrating the DInSAR data with the GPS displacements, calculated through PPP (Precise Point Positioning) processing by the GIPSY-OASIS package on a daily basis over the same InSAR time window. This processing is independent and widely different from the high-rate (HR) one, not only due to the different observation times but also being performed on daily standard RINEX (Receiver INdependent EXchange) data, decimated from the original 1 to 30 s rate, processed in static mode to achieve a unique, precise and not-moving solution for each day. 3D GPS and 1D/LOS DInSAR displacements were integrated by the SISTEM algorithm [26], a software that allows the complete integration of different ground deformation datasets in order to calculate, point-by-point with the DInSAR ground resolution, the 3D components of the ground motion over the 12-day period covered by DInSAR data. In this way, an integrated map can be produced, similar to an interferogram for spatial resolution but containing, for each pixel, the 3D motion of the ground (instead of the LOS component of the DInSAR image).

The main source of error of the DInSAR measurements is due to atmosphere noise, especially related to the strong topography at Etna (often more than a fringe from base to summit, see [27,28] and references therein). For this reason a preliminary step, before their use, is the validation-correction of the DInSAR data with GPS and/or other independent measures. The SISTEM algorithm not only integrates but also simultaneously validates the DInSAR data with GPS by taking advantage of the positive features of both these techniques, i.e., the high spatial resolution of the DInSAR and the 3-D measurements and subcentimeter accuracy level of the GPS. This integrated and cross-validated information is able to give a more reliable interpretation of the geophysical phenomena producing ground deformations.

The general pattern evidenced by DInSAR data (Figure 7) is mostly coherent with the deformation highlighted by the 12-day GPS displacement pattern (thick arrows in Figure 8), both evidencing, even with different resolution and completeness of information (1D but high-resolution for InSAR and 3D but low-resolution for GPS), a contraction of the volcano. The integration of both datasets (thin arrows and colors in Figure 8) images a general deflation of the edifice well, especially evident on its western and southern slopes. The match between the integrated and original observed GPS and DInSAR datasets has been checked and validated for looking at possible artifacts. Results of the validation process evidence a mean difference between observed GPS displacements and SISTEM displacements at GPS points of about 1–1.5 mm (abundantly within the GPS errors) and differences within 5 mm for most of the DInSAR interferogram (figures GPS vs. SISTEM and SAR vs. SISTEM provided in the Supplementary Materials). The main deflation seems to affect the upper part of the volcano, above 1700–1800 m a.s.l. (in agreement with the interferogram, see Figure 7), which shows an evident radial contraction and a subsidence of the ground.

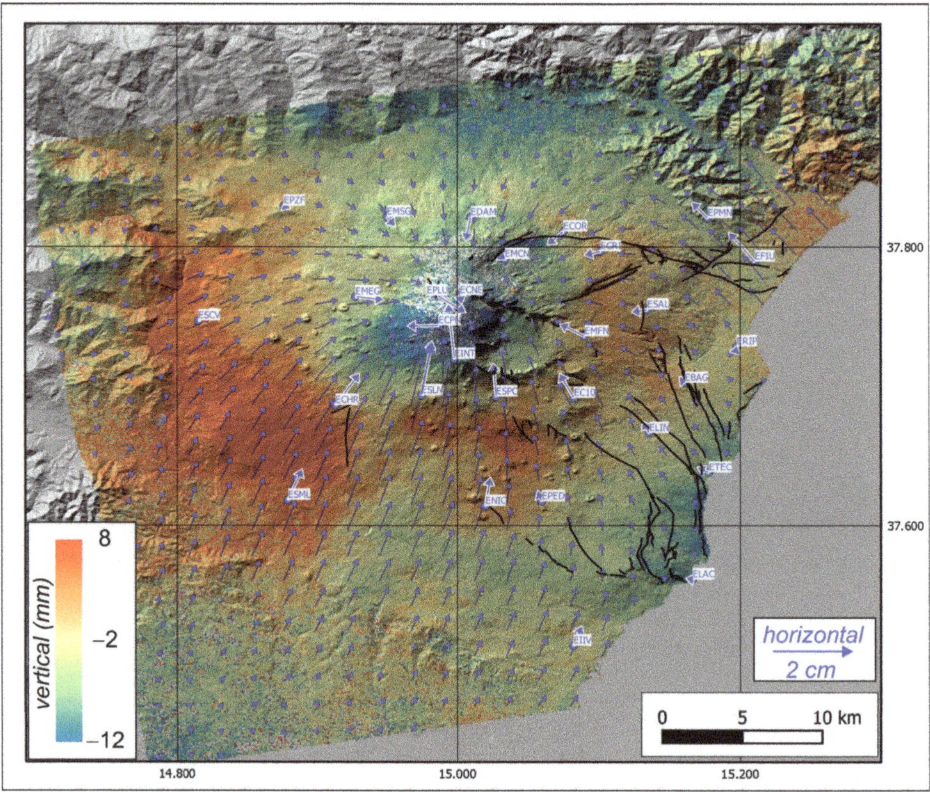

Figure 8. 3D displacement map obtained by SISTEM algorithm over the December 2 to 14 period. Integrated horizontal displacements are shown by (sub-sampled) blue vectors, while the vectors showing the GPS-only measured displacements are evidenced by thicker arrows with white borders. The color map shows the integrated vertical displacements, according to the color scale on the bottom left.

A slight uplift is visible at a lower altitude, on the western and southern slopes. This particular feature lets us suppose that two dynamics occurred in the 12-day interval encompassing the four fire fountains sequence: the rapid depletion of the feeding system, related to the four eruptions and leading to the summit deflation with, probably, a simultaneous pressurization of a deeper source, producing a wider uplift, visible at the volcano periphery, where it is not masked by the local deflation. These wider and, reasonably, deeper dynamics cannot be investigated here, due to the superimposition of the much stronger deflation; it needs more data, covering also longer and different time windows, to be constrained and it is beyond the scope of this study.

4. Modeling

We used high-rate measurements from both GPS baselines and tilt variations to estimate the causal sources of each paroxysm and, then, the integrated 3D SISTEM ground deformation map to search for a cumulative source feeding the entire sequence. In this way we exploited the availability and potential of HR data showing the continuous and rapid deformation, even with a non-optimal signal-to-noise ratio, and the powerful integrated map, in terms of both spatial resolution and quality of the signal. This allowed us to identify the single-shot sources feeding each paroxysm and the entire feeding system involved during the whole sequence.

Among all the possible HR baselines, we chose the baselines that showed clear patterns, avoiding too-noisy baselines and redundancy (using a single GNSS site for at most two baselines).

To calculate the displacements and their associated uncertainties, we fitted the baseline time series with a piecewise linear regression line. The variations were calculated from the segments at the eruptive periods.

Similarly, the tilt variations in the same periods were considered in the joint inversion.

The analytical model adopted to fit the geodetic data was the finite spherical source model [29] and we used the GAME software [30] to invert the data, applying a hybrid approach of genetic algorithm and pattern search [31]. The objective function was the sum of the squared model residues weighted with their associated uncertainties. The model parameter uncertainties were estimated by using a Jackknife approach [30,32].

The resulting sources show similar depths and coherent horizontal positions (within the area of summit craters) as shown in Table 1. The location of the source was very stable for the three paroxysms; taking into account the differences among the three solutions, the results clearly define a volume located at around 1.8 km depth beneath the summit craters area. The model fitting with the measured deformation data degrades as the energy of the eruption decreases; it is good for the first event, becoming just sufficient for the third event (see Figure 9).

In fact, the first two, stronger, paroxysms are those with the strongest signal-to-noise ratio and this allows a better match between predicted and measured deformation. The third one shows a less-fitting modeled deformation, especially on baselines 6–4 and 1–7 (Figure 9), due to a lower signal-to-noise ratio in the deformation data.

The last fountain did not show enough deformation to be coherently measured by a sufficient number of considered GPS baselines. Thus, we avoided modeling the source. The model's fit with tilt data is shown in Table 2.

Table 1. Parameters of the single-shot sources for the first three paroxysms, from HR GNSS and tilt data, together with the associated uncertainties at 2-sigma.

	Paroxysm 1	**Paroxysm 2**	**Paroxysm 3**
UTM Easting (m)	498,921 ± 192	498,694 ± 95	499,017 ± 150
UTM Northing (m)	4,177,360 ± 166	4,177,110 ± 105	4,179,300 ± 172
Depth b.s.l. (m)	1876 ± 450	1840 ± 632	1763 ± 1032
Volume Change (M m^3)	−1.4 ± 0.2	−1.0 ± 0.2	−0.9 ± 0.2

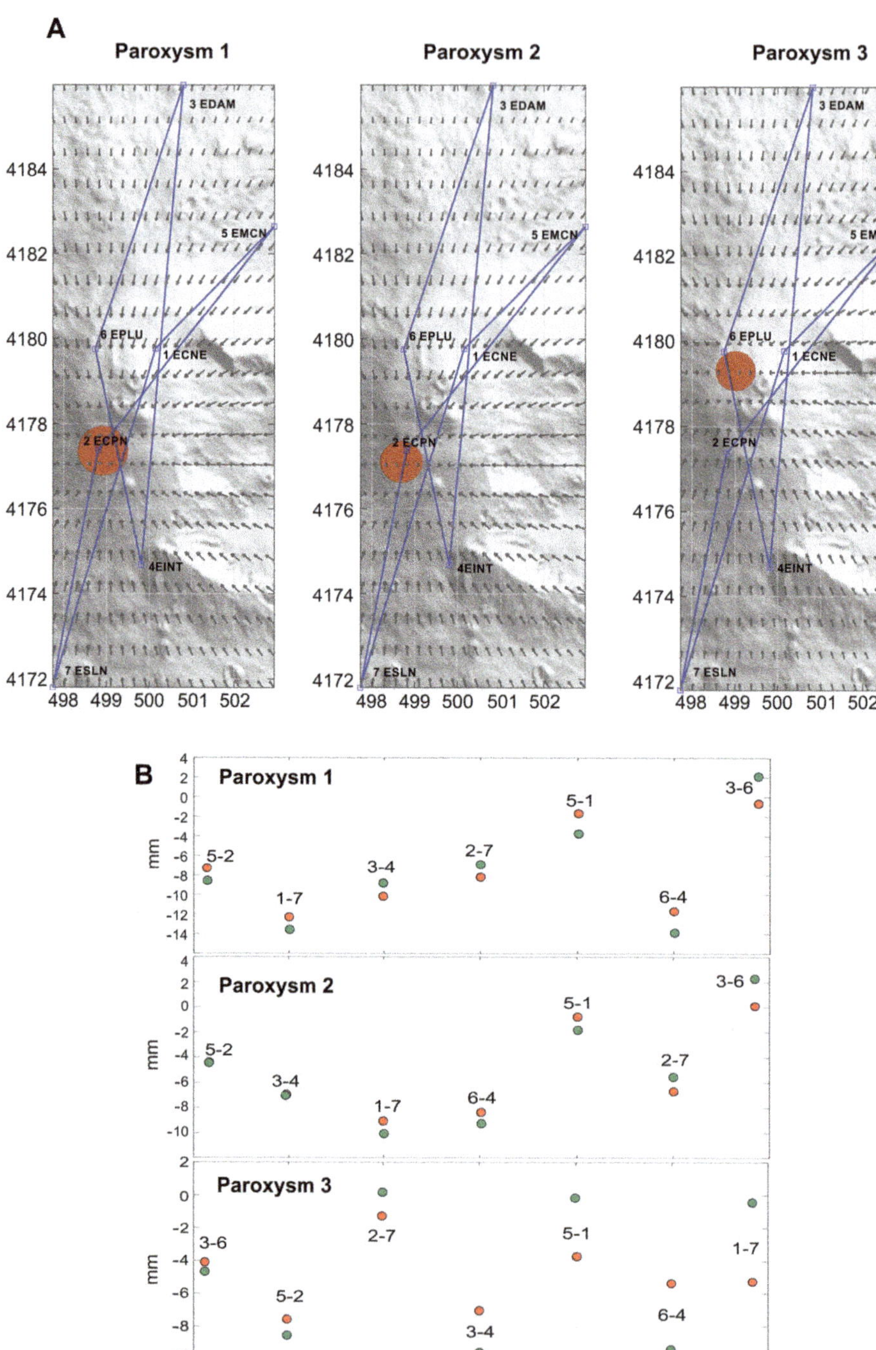

Figure 9. Model from the HR data relevant to the first three paroxysms. (**A**) Red circle indicates the position of the source, gray arrows indicate the expected displacements and blue lines indicate the inverted baselines; (**B**) model fitting: green points indicate the measured baseline variations, red points indicate the model predictions. The baselines are sorted in descending order of the error (misfit).

Table 2. Measured tilt components vs. model predictions for the 3 paroxysms. Tilt data are reported in microrads.

| | Paroxysm 1 | | | | Paroxysm 2 | | | | Paroxysm 3 | | | |
| | Measured | | Predicted | | Measured | | Predicted | | Measured | | Predicted | |
	x	y	x	y	x	y	x	y	x	y	x	y
PLC	0.27	1.01	−0.41	2.35	0.19	0.82	−0.05	1.95	0.05	0.50	0.98	0.17
CDV	0.67	−0.22	0.44	−0.49	0.60	−0.20	0.35	−0.35	0.38	−0.08	0.20	−0.26
CBD	0.69	0.21	0.20	0.10	0.54	0.26	0.13	0.07	0.35	0.10	0.29	0.05
MNR	0.21	0.45	0.14	0.37	0.21	0.45	0.10	0.24	0.11	0.23	0.58	0.75
PDN	0.44	0.44	1.88	1.95	0.27	0.29	1.19	1.01	0.08	0.10	0.92	0.08
DAM	0.31	1.17	0.04	0.27	0.13	0.89	0.03	0.18	0.07	0.50	0.31	0.78
MGL	−0.72	−0.37	−0.25	−0.08	−0.51	−0.35	−0.21	−0.06	−0.36	−0.18	−0.32	−0.22

In order to model the source feeding the entire sequence, that should represent the average location of the origin of the magma drained during the four paroxysms, we analyzed the longer-term (12 days) but lower-noise data coming from GNSS and DInSAR. The 3D deformation pattern depicted by integrating GNSS and DInSAR data from December 2 to 14 clearly defines a radial contraction of the entire edifice, accompanied by a general subsidence centered on the summit part, indicating an evident deflation of the edifice. A circular slight uplift at a lower altitude permits hypothesizing that a deep source was pressurizing, able to produce a wider smooth effect, underlying the more evident and stronger deflation. However, this uplift was not accompanied by radial horizontal expansion, the dominating deformation being the centripetal direction of horizontal displacements imputable to the deflation. For this reason, the deep pressurizing source cannot be constrained by a clear deformation and we concentrated our efforts on modeling the deflating source.

In order to define the source of the observed cumulative deflation, we inverted the 3D SISTEM integrated data using a pressure source model, as formulated by [33], and implemented it into the dMODELS Matlab® software package [34]. To search the minimum of the residuals, we used an optimization routine based on the genetic algorithms approach, as modified by [5], using the same procedure described in [23] by considering also the volcano topography.

The goodness of the SISTEM integration, depends on the number of GPS points and on the geometry of the GPS network, and this issue was largely investigated in [26].

Before using the integrated data, the deformation calculated by the integration has been cchecked and validated at all corresponding GPS points, to ensure that it reproduces what really observed by GPS; we also revert the SISTEM output into LOS displacements for calculating the fit with the original DInSAR data (see supplementary material). Once being sure that SISTEM integration does not introduce artifacts and unreal signals, we think that inverting the integrated and validated data provides a better-constrained model than inverting only the original GPS and DInSAR data. In this way, we gain millions of points with 3D information instead of few tens with 3D and millions with only 1D.

The solution converged to a final fitness value of 75% with an average misfit of 1.4 and 4 mm for the horizontal and vertical components, respectively. The distribution of the residuals (Figure 10) shows a general good fit of the expected deformation with the observed one all over the volcano. Larger residuals can be observed on the vertical component, where the uplift at a lower altitude remains unmodeled, while the subsidence due to the deflation is well matched by the model.

The general solution for a prolate spheroid depends on eight parameters: the dimensionless pressure change $\Delta P/\mu$; the semimajor axis; the geometric aspect ratio A between the length of the semimajor axis a and that of the two semiminor axes b; the source location (x,y,z); the dip angle θ (measured from the free surface) and the azimuth angle φ (measured clockwise from the positive North direction). In addition, we calculated the volume change of the spheroid, according to [35].

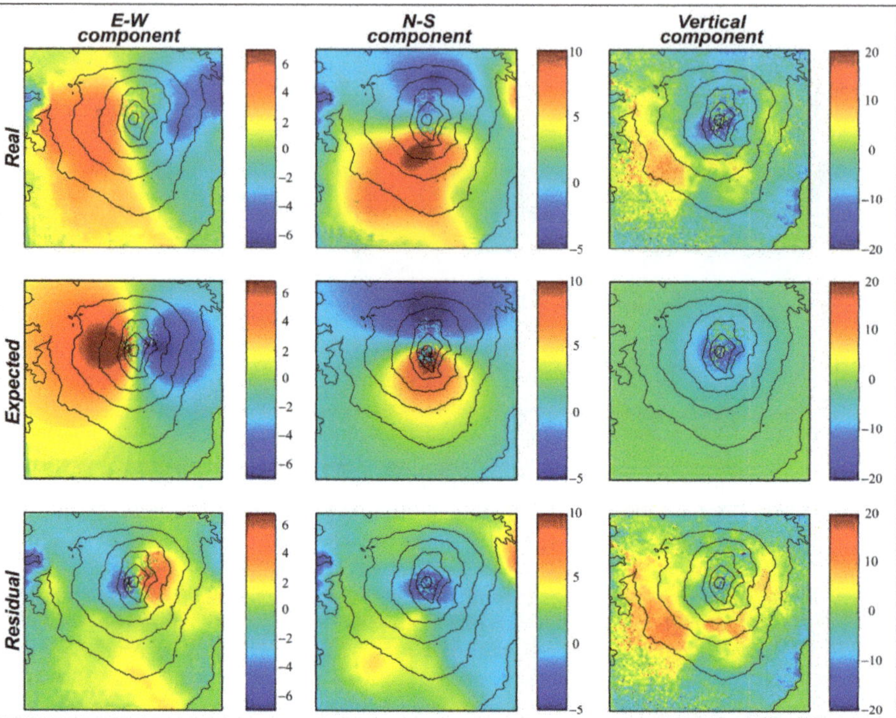

Figure 10. SISTEM inversion results. In the first row the real East–West, North–South and Vertical components are reported; in the second and third row the corresponding Expected and Residual are reported, respectively; all the values are in mm.

The solution converged to an almost vertically dipping (78°) prolate spheroid, N–S oriented (N185E), located at about 6 km beneath the summit craters, losing about 6.5 million cubic meters during the entire eruptive cycle (Table 3), which is close to the volume (7 million cubic meters) estimated by [2]. This location should represent the original source of magma, that was gradually depleted to feed the shallower single-shot sources, where the gas exsolved and pressurized to charge each paroxysm.

Table 3. Parameters of the source modeled from SISTEM integrated deformation data.

Parameter	Value
UTM Easting (m)	500,055 ± 59
UTM Northing (m)	4,179,769 ± 19
Depth b.s.l. (m)	6530 ± 22
Radius (m)	500 ± 0 (fixed)
A (b/a)	0.238 ± 0.01
$\Delta P/\mu$	−0.18 ± 0.03
Dip Angle	78 ± 1°
Azimuth	185 ± 2°
Volume Change (M m^3)	−6.5

5. Discussion

HR data coming from tiltmeters and GNSS networks on Mt. Etna allowed us to investigate in detail the very rapid deformation accompanying each of the four paroxysmal episodes of the Voragine crater at Mt. Etna. From the time-series analysis it is evident that the four paroxysms produced decreasing ground displacements, both on GNSS and tilt signals, confirming the decreasing energy and volumes emitted, as reported by [2]. By inverting the ground deformation data recorded during the first three events, the location of the modeled sources appears quite stable. Indeed, although the sources were estimated separately in each of the considered periods, the coherence in the retrieved positions suggests the presence of a single depletion volume; in fact considering the dispersion of the three solutions, all of them fall in a small volume located beneath the summit craters at a depth of about 5 km below the summit craters area. The involved volumes range from 0.9 to 1.4×10^6 m^3. The total volume of the first three paroxysms, coming from the modeled sources by HR data, is about 3.3×10^6 m^3. Even considering a hypothetical volume of the fourth one, that should not exceed that of the third, the total volume is lower than the volume erupted, as calculated by [2] for the entire sequence (around 7×10^6 m^3). This discrepancy could be related to the noisy data with respect to the low deformation signal and/or to the not completely elastic behavior of the medium close to the surface and around the summit craters and feeding system.

We should also consider that the shallow single-shot source mainly represents a level where part of the magma, coming from depth, is temporarily stored for a few hours before each paroxysm. Thus, the deflation does not necessarily represent the total volume erupted from this source but only a partial variation of volume, the magma being mainly in transit there.

The cumulated ground deformation revealed by more precise daily GNSS solution and 12-days Sentinel-1 SAR interferogram depicts an overall wide deflation and contraction of the entire edifice, suggesting a significant depletion of a deeper source, with respect to those resulting from HR data inversions. The integration of both datasets makes the ground deformation pattern even more evident and precise and allows a well-constrained inversion to search for the source feeding the eruptive sequence. The modeled source is a vertical and N–S oriented prolate spheroid, located at a higher depth of almost 6 km b.s.l., always beneath the summit craters area. The volume lost by this source is about 6×10^6 m^3, much more similar to the 7.1×10^6 m^3 calculated by [2] for the entire sequence, with respect to the single-shot sources. This seems to confirm the role of the shallower source as a local and temporary storage of magma in transit and coming, instead, from this middle-depth source at 6 km b.s.l. The DInSAR and integrated data also reveal a slight uplift at lower altitudes on the flanks of the volcano; even though proving impossible to be well constrained and modeled, this information enabled hypothesizing a further and deeper source that was pressurizing during the entire period. As stated before, this eventual deeper source needs more data to be constrained, over larger and different time windows, and is beyond the scope of this investigation.

Considering this pulsating behavior of the shallow source, we can postulate that the shallower source (~1.5–2 km b.s.l.) slowly accumulated, before each paroxysm, the same volume of magma lost during each episode (as resulting from HR data inversions). We can use the volume changes modeled and time intervals occurring between two consecutive paroxysms to infer the magma feeding rate from the middle depth (~6 km b.s.l.) to the shallower source, during each quiet phase. We can estimate that the shallow source charged at a rate of about 10 m^3/s before the second episode, 25 m^3/s before the third one and about 15 m^3/s before the last lava fountain (considering a volume not greater than the third one). This gives an average *recharging rate* of around 16 m^3/s of magma ascent during the quiet inter-paroxysm recharging phases. If we consider instead the total volume depleted from the middle-depth source (6.5×10^6 m^3), we can estimate an *overall supply rate* of about 29 m^3/s of magma that moved from 6 km b.s.l. to the surface during the entire 2.6-days period, or 31 m^3/s if we consider the total volume (7.1×10^6 m^3) reported by [2]. This mean value (29 to 31 m^3/s) averages a rate that actually oscillated from the *recharging rate* of ~16 m^3/s during closed-vent and non-eruptive conditions (magma moving from 6 to 1.5–2 km b.s.l), to the peak *eruptive discharge rate* of more than 300 m^3/s during the paroxysms (considering a mean volume of about 1.8×10^6 m^3 erupted at each paroxysm in about 90 min).

We can perform a very preliminary comparison of this analysis with the last and ongoing (at the time of writing) sequence of paroxysms at SEC (started on 16 February 2021). There is still no precise estimation of the erupted volumes, so we can roughly estimate a total volume of about 45 to 60 $\times$ 10^6 m^3 erupted during 15 strong episodes occurring from 16 February to 19 March 2021 (based on the very first estimation of the volumes from INGV bulletins available at https://www.ct.ingv.it/index.php/monitoraggio-e-sorveglianza/prodotti-del-monitoraggio/bollettini-settimanali-multidisciplinari, accessed on 5 April 2021). Starting from this preliminary information, an *overall supply rate* of around 17 to 22 m^3/s can be estimated over the entire month. This rate is comparable to what happened in 2015 and analyzed here, leading to the supposition that the supply mechanism, geometry and rates hypothesized for the 2015 paroxysms can be valid generally for other similar eruptive sequences on Mt. Etna.

6. Conclusions

The coupling of high-rate, even if noisy, and precise, albeit if low-rate, data allowed us to perform a comprehensive investigation of the dynamics leading to and enabling very rapid magma movement and pressure accumulation and discharge, through an impressive sequence of very violent eruptions. The overall picture coming from the different rates data analyses constrains the feeding system underlying such a quick and violent sequence of eruptive episodes (well summarized in Figure 11), allowing a rapid magma upraise and discharge from the summit crater, located at about 3300 m a.s.l. at Mt. Etna. In this framework, there is a deep, and not well-constrained here, pressurizing source that provides gas-rich magma to a middle-depth reservoir, located at about 6 km depth. This is the reservoir that gradually feeds the shallower system at about 1.5–2 km b.s.l., where the upper part of the magma batch temporarily stops and vesiculates, charging and producing each paroxysm when the pressure, mainly due to the exsolution of gas, exceeds the containment pressure of the hosting rocks inside the volcano.

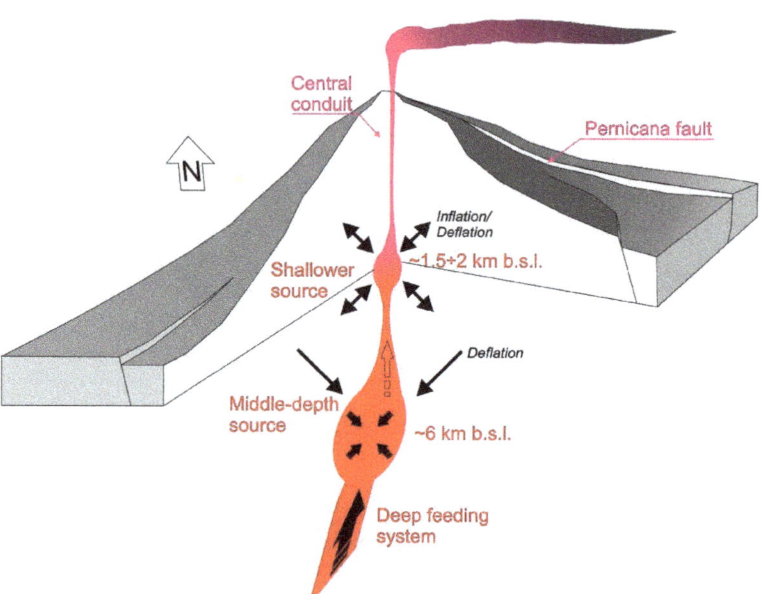

Figure 11. Sketch of the feeding system resulting from HR and integrated ground deformation modeling. A deep and fairly continuous magma upraise feeds a middle-depth storage level. During the eruptive sequence, this storage is depleted to feed a shallower reservoir that works as a temporary storage, charging to feed each single-shot paroxysm.

The combined mechanism of two storage levels at different depths depicts the possible feeding system of such high-discharge rate events, with a middle-depth storage of gas-rich magma that charges, more or less continuously, at a *recharging rate* of about 15–20 m^3/s, and a shallower level where magma stops temporarily, accumulating pressure due to the gas exsolution. When the pressure exceeds a threshold determined by lithostatic pressure and structural conditions of the shallower feeding system, the paroxysm starts, draining not only the magma accumulated in this shallower source but also part of that in the rest of the feeding system at an *eruptive discharge rate* of more than 300 m^3/s. When the gas pressure is discharged below the threshold, the paroxysm ends and the shallower source starts to be re-charged again. The *overall supply rate* of the entire sequence is about 29–31 m^3/s, which is similar to that estimated for the most recent sequence at SEC (February–March 2021). Thus, the model suggests a machine-gun-like mechanism, comprising a middle-depth reservoir continuously charging a shallower one that works as a firing chamber triggering each single-shot paroxysm. This mechanism, and the overall supply rates needed to feed this dynamic, could represent a general conceptual model for similar events at Etna and at all volcanoes.

Supplementary Materials: The following are available online at https://www.mdpi.com/article/10.3390/app11104630/s1, XLS files refer to the tilt variations measured from December 2 to 7 at the eight stations here considered. In the other three XLS files, the tilt measured for the first three paroxysms are summarized. Each file lists the station name, its coordinates and its tilt components (in microradians). The baseline variations are stored in a data structure within the MAT file. The structure includes the station names and locations, and a table with the baseline variations estimated for each period of the piecewise linear regression. 3D_SISTEM_output.zip: zipfile containing Ue_SISTEM.asc, Un_SISTEM.asc, Uu_SISTEM.asc SISTEM 3D Output (East, North and Up components of GPS and DInSAR integrated deformation, respectively) in ESRI- ASCII grid format, Word Geodetic System 1984 UTM, zone 33N (IGNF:UTM33W84), reference system (m) and a README.TXT file describing these files. GPS.csv: GPS displacement relevant to 2–14 December 2015 timespan.

Author Contributions: Conceptualization, A.B.; validation, A.B., F.C., S.G. and F.G.; data curation, F.C., S.G. and F.G.; methodology, F.C. and F.G.; formal analysis, F.C. and F.G.; writing—original draft preparation, A.B.; writing—review and editing, A.B., F.C., S.G. and F.G.; supervision, A.B. and S.G.; investigation, A.B. All authors have read and agreed to the published version of the manuscript.

Funding: This research received no external funding.

Institutional Review Board Statement: Not applicable.

Informed Consent Statement: Not applicable.

Data Availability Statement: We provide, as Supplementary Material, all the data processed and analyzed in this work.

Acknowledgments: The interferometric data are provided in the framework of the GEO GNSL initiative. Copernicus Sentinel-1 data (2015) are available at the Copernicus Open Access Hub (https://scihub.copernicus.eu, accessed on 1 January 2016). Pleiades data have been provided in the framework of the MED-SUV project. The authors also thank the INGV technical group of the GPS and Tilt permanent networks to ensure the regular working of the stations at Mount Etna. Scientifict activities have been carried out also in the framevork of WP1 of the INGV IMPACT Project (CUP D59C19000130005).

Conflicts of Interest: The authors declare no conflict of interest.

References

1. Acocella, V.; Neri, M.; Behncke, B.; Bonforte, A.; Del Negro, C.; Ganci, G. Why Does a Mature Volcano Need New Vents? The Case of the New Southeast Crater at Etna. *Front. Earth Sci.* **2016**, *4*. [CrossRef]
2. Corsaro, R.A.; Andronico, D.; Behncke, B.; Branca, S.; Caltabiano, T.; Ciancitto, F.; Cristaldi, A.; De Beni, E.; La Spina, A.; Lodato, L.; et al. Monitoring the December 2015 summit eruptions of Mt. Etna (Italy): Implications on eruptive dynamics. *J. Volcanol. Geotherm. Res.* **2017**, *341*, 53–69. [CrossRef]

3. Bonaccorso, A.; Calvari, S. A new approach to investigate an eruptive paroxysmal sequence using camera and strainmeter networks: Lessons from the 3–5 December 2015 activity at Etna volcano. *Earth Planet. Sci. Lett.* **2017**, *475*, 231–241. [CrossRef]

4. Alparone, S.; Andronico, D.; Lodato, L.; Sgroi, T. Relationship between tremor and volcanic activity during the Southeast Crater eruption on Mount Etna in early 2000. *J. Geophys. Res. Space Phys.* **2003**, *108*, 2241. [CrossRef]

5. Nunnari, G.; Puglisi, G.; Guglielmino, F. Inversion of SAR data in active volcanic areas by optimization techniques. *Nonlinear Process. Geophys.* **2005**, *12*, 863–870. [CrossRef]

6. Calvari, S.; Neri, M.; Pinkerton, H. Effusion rate estimations during the 1999 summit eruption on Mount Etna, and growth of two distinct lava flow fields. *J. Volcanol. Geotherm. Res.* **2003**, *119*, 107–123. [CrossRef]

7. Ferro, A.; Gambino, S.; Panepinto, S.; Falzone, G.; Laudani, G.; Ducarme, B. High precision tilt observation at Mt. Etna Volcano, Italy. *Acta Geophysica* **2011**, *59*, 618–632. [CrossRef]

8. Gambino, S.; Falzone, G.; Ferro, A.; Laudani, G. Volcanic processes detected by tiltmeters: A review of experience on Sicilian volcanoes. *J. Volcanol. Geotherm. Res.* **2014**, *271*, 43–54. [CrossRef]

9. Gambino, S.; Cammarata, L. Tilt measurements on volcanoes: More than a hundred years of recordings. *Ital. J. Geosci.* **2017**, *136*, 275–295. [CrossRef]

10. Puglisi, G.; Briole, P.; Bonforte, A. Twelve years of ground deformation studies on Mt. Etna volcano based on GPS surveys. In *Mt. Etna: Volcano Laboratory*; Bonaccorso, A., Calvari, S., Coltelli, M., Del Negro, C., Falsaperla, S., Eds.; Geophysical Monograph Series 143; AGU: Washington, DC, USA, 2004; pp. 321–341.

11. Bonforte, A.; Guglielmino, F.; Puglisi, G. Interaction between magma intrusion and flank dynamics at Mt. Etna in 2008, imaged by integrated dense GPS and DInSAR data. *Geochem. Geophys. Geosyst.* **2013**, *14*, 2818–2835. [CrossRef]

12. Bonaccorso, A.; Bonforte, A.; Gambino, S. Twenty-five years of continuous borehole tilt and vertical displacement data at Mount Etna: Insights on long-term volcanic dynamics. *Geophys. Res. Lett.* **2015**, *42*, 10,222–10,229. [CrossRef]

13. Bruno, V.; Mattia, M.; Aloisi, M.; Palano, M.; Cannavò, F.; Holt, W.E. Ground deformations and volcanic processes as imaged by CGPS data at Mt. Etna (Italy) between 2003 and 2008. *J. Geophys. Res. Solid Earth* **2012**, *117*. [CrossRef]

14. Bock, Y.; Nikolaidis, R.M.; De Jonge, P.J.; Bevis, M. Instantaneous geodetic positioning at medium distances with the Global Positioning System. *J. Geophys. Res. Space Phys.* **2000**, *105*, 28223–28253. [CrossRef]

15. Zumberge, J.F.; Heflin, M.B.; Jefferson, D.C.; Watkins, M.M.; Webb, F.H. Precise point positioning for the efficient and robust analysis of GPS data from large networks. *J. Geophys. Res. Solid Earth* **1997**, *102*, 5005–5017. [CrossRef]

16. Chen, G. GPS Kinematic Positioning for the Airborne Laser Altimetry at Long Valley, California. Ph.D. Thesis, Department of Earth, Atmospheric and Planetary Sciences, Massachusetts Institute of Technology, Cambridge, MA, USA, 1998.

17. Herring, T.; King, R.W.; McClusky, S. *GAMIT Reference Manual*; Release 10.4 214; Massachusetts Institute of Technology: Cambridge, MA, USA, 2011.

18. Ganas, A.; Cannavo, F.; Chousianitis, K.; Kassaras, I.; Drakatos, G. Displacements recorded on continuous gps stations following the 2014 M6 cephalonia (Greece) earthquakes: Dynamic characteristics and kinematic implications. *Acta Geodyn. Geomater.* **2015**, *12*, 5–19. [CrossRef]

19. Huber, P.J. *Robust Statistics*; Springer: Berlin/Heidelberg, Germany, 2011.

20. Bonforte, A.; Guglielmino, F.; Coltelli, M.; Ferretti, A.; Puglisi, G. Structural assessment of Mount Etna volcano from Permanent Scatterers analysis. *Geochem. Geophys. Geosyst.* **2011**, *12*. [CrossRef]

21. Alparone, S.; Cocina, O.; Gambino, S.; Mostaccio, A.; Spampinato, S.; Tuvè, T.; Ursino, A. Seismological features of the Pernicana–Provenzana Fault System (Mt. Etna, Italy) and implications for the dynamics of northeastern flank of the volcano. *J. Volcanol. Geotherm. Res.* **2013**, *251*, 16–26. [CrossRef]

22. Bonforte, A.; Branca, S.; Palano, M. Geometric and kinematic variations along the active Pernicana fault: Implication for the dynamics of Mount Etna NE flank (Italy). *J. Volcanol. Geotherm. Res.* **2007**, *160*, 210–222. [CrossRef]

23. Guglielmino, F.; Bignami, C.; Bonforte, A.; Briole, P.; Obrizzo, F.; Puglisi, G.; Stramondo, S.; Wegmuller, U. Analysis of satellite and in situ ground deformation data integrated by the SISTEM approach: The April 3, 2010 earthquake along the Pernicana fault (Mt. Etna—Italy) case study. *Earth Planet. Sci. Lett.* **2011**, *312*, 327–336. [CrossRef]

24. Alparone, S.; Barberi, G.; Bonforte, A.; Maiolino, V.; Ursino, A. Evidence of multiple strain fields beneath the eastern flank of Mt. Etna volcano (Sicily, Italy) deduced from seismic and geodetic data during 2003–2004. *Bull. Volcanol.* **2011**, *73*, 869–885. [CrossRef]

25. Alparone, S.; Bonaccorso, A.; Bonforte, A.; Currenti, G. Long-term stress-strain analysis of volcano flank instability: The eastern sector of Etna from 1980 to 2012. *J. Geophys. Res. Solid Earth* **2013**, *118*, 5098–5108. [CrossRef]

26. Guglielmino, F.; Nunnari, G.; Puglisi, G.; Spata, A. Simultaneous and Integrated Strain Tensor Estimation from Geodetic and Satellite Deformation Measurements to Obtain Three-Dimensional Displacement Maps. *IEEE Trans. Geosci. Remote. Sens.* **2011**, *49*, 1815–1826. [CrossRef]

27. Bonforte, A.; Ferretti, A.; Prati, C.; Puglisi, G.; Rocca, F. Calibration of atmospheric effects on SAR interferograms by GPS and local atmosphere models: First results. *J. Atmos. Solar-Terr. Phys.* **2001**, *63*, 1343–1357. [CrossRef]

28. Bruno, V.; Aloisi, M.; Bonforte, A.; Immè, G.; Puglisi, G. Atmospheric anomalies over Mt. Etna using GPS signal delays and tomography of radio waves velocities. *Ann. Geophys.* **2007**, *50*, 267–282.

29. McTigue, D.F. Elastic stress and deformation near a finite spherical magma body: Resolution of the point source paradox. *J. Geophys. Res. Solid Earth* **1987**, *92*, 12931–12940. [CrossRef]

30. Cannavò, F. A new user-friendly tool for rapid modelling of ground deformation. *Comput. Geosci.* **2019**, *128*, 60–69. [CrossRef]

31. Audet, C.; Dennis, J.E., Jr. Analysis of generalized pattern searches. *SIAM J. Optim.* **2002**, *13*, 889–903. [CrossRef]
32. Efron, B. Nonparametric estimates of standard error: The jackknife, the bootstrap and other methods. *Biometrika* **1981**, *68*, 589–599. [CrossRef]
33. Yang, X.-M.; Davis, P.M.; Dieterich, J.H. Deformation from inflation of a dipping finite prolate spheroid in an elastic half-space as a model for volcanic stressing. *J. Geophys. Res. Solid Earth* **1988**, *93*, 4249–4257. [CrossRef]
34. Battaglia, M.; Cervelli, P.F.; Murray, J.R. dMODELS: A MATLAB software package for modeling crustal deformation near active faults and volcanic centers. *J. Volcanol. Geotherm. Res.* **2013**, *254*, 1–4. [CrossRef]
35. Amoruso, A.; Crescentini, L. Shape and volume change of pressurized ellipsoidal cavities from deformation and seismic data. *J. Geophys. Res. Solid Earth* **2009**, *114*, 02210. [CrossRef]

*applied
sciences*

Article

Imaging Top of Volcanic Mounds Using Seismic Time- and Depth-Domain Data Processing

Woohyun Son, Snons Cheong *, Changyoon Lee and Moohee Kang

Petroleum & Marine Division, Korea Institute of Geoscience and Mineral Resources, Daejeon 34132, Korea;
whson@kigam.re.kr (W.S.); chngynlee@kigam.re.kr (C.L.); karl@kigam.re.kr (M.K.)
* Correspondence: snons@kigam.re.kr; Tel.: +82-42-868-3152

Abstract: A seismic survey identified a basalt flow that could consist of cap rock of CO_2 storage beneath saline aquifer sediment in the Southern Continental Shelf of Korea. To determine the precise depth of the basalt flow, specific depth-domain data processing of migration velocity analysis (MVA) was applied to the seismic survey data. The accurate depth measurement of a target structure provides crucial information when storing and stabilizing injected CO_2 beneath basalt cap rock. Strong reflections of seismic amplitude at the volcanic mounds were adjusted from the time domain to the exact depth domain by the iterated velocity using MVA. The confidence of the updated velocity was verified by the horizontal alignment of seismic events sorted according to their common reflection point (CRP). The depth difference in volcanic mounds before and after MVA application ranged from 32.5 to 60 m along the vertical axis, showing the eruption shape on the strong-amplitude contour map in detail. The eruption shape of the top of volcanic mounds was verified with spatial continuity in 3D geological interpretation. The presented results provide suitable information that can be used to locate drilling sites and to prepare CO_2 injection. The geological model obtained from both time- and depth-domain processing can significantly influence the calculation of the storage volume and can be useful for history matching studies.

Keywords: volcanic mounds; seismic time and depth processing; MVA; CO_2 storage

Citation: Son, W.; Cheong, S.; Lee, C.; Kang, M. Imaging Top of Volcanic Mounds Using Seismic Time- and Depth-Domain Data Processing. *Appl. Sci.* **2021**, *11*, 4244. https://doi.org/10.3390/app11094244

Academic Editor: Alessandro Bonforte

Received: 22 March 2021
Accepted: 4 May 2021
Published: 7 May 2021

1. Introduction

Areas affected by volcanic eruptions such as basalt lava flows or fissure vents are frequently found across the Earth's surface, arousing considerable scientific interest [1]. The volcanic subsurface can be explored through the use of geophysical methodologies, e.g., gravity, magnetic and seismic surveys [2–4]. In the seismic exploration data, the volcanic area shows strong amplitude events due to the relatively high contrast of density and reflectivity [5]. Abrupt collapses, unexpected faults and subsidence are other common characteristics of magmatic intrusions resulting from volcanic activity [6]. Aspects of the reflection signal's diversity present challenges in seismic data processing, particularly in the estimation of velocity information [6–8]. Since the seismic data related to volcanic activity have less commercial usefulness than the exploration data for hydrocarbon development, the complicated data-processing procedure is often not applied. In addition, representative research cases in which the velocity was obtained by processing seismic data have used geophone-recorded data, which contain accurate depth information [7,8].

The increase in the scientific and/or engineering curiosity regarding volcanic regions means that additional velocity-specific seismic data processing is required. Carbon capture and storage (CCS) projects have led to expanded efforts to assess the storage, from depleted reservoirs to saline aquifers with upper volcanic cap rocks [9,10]. Saline aquifers have the advantage of large storage volume but require a sealing structure in order to safely store CO_2 [10,11]. Basalt flow, due to its rock-physical properties, may be a suitable cap rock for the storing CO_2 in the aquifer beneath the flow. Moreover, the mineralization of carbon can prevent the migration of CO_2, which leads to favorable stabilization for long-term storage [12]. To use basalt flow as the cap rock for CO_2 storage, accurate verification of the fracture zone and deformation should be carried out in both the time and depth domains [13]. Optimized seismic data processing for basalt flow can provide an accurate understanding of the possible fracture zone due to velocity anomalies and possible deformations by subtle depth conversion.

In the Southern Continental Shelf of Korea (SCSK), there is an independent structure formed by volcanic events, showing the potential of CCS [14,15]. Two-dimensional seismic surveys were executed in 2018 across the areas of the SCSK to calculate the storage capacity of CO_2 for the saline aquifer. Determination of the exact depth of the target is important when drilling injections and monitoring wells [16]. In the seismic sections, by applying only time-domain processing, basalt flow shows high reflectivity contrast, which might lead to errors during the geological interpretation. The boundary and lateral expansion of basalt flow are not easy to analyze in terms of potential cap rocks for CO_2 storage.

To enhance the seismic sections' resolution, we applied both time-domain and depth-domain processing using the MVA technique [17,18]. MVA is able to determine the velocity with high accuracy by considering subtle changes in spatial velocities. We demonstrated the iterated velocity and determined the accurate velocity from the flatness of the collected CRP gathers. In order to confirm the effectiveness of our processing, amplitude contour and root mean square (RMS) maps derived before and after MVA application were compared. If the precise boundary and depth of volcanic mounds can be defined, CO_2 storage capacity can be calculated with confidence.

2. Structural and Geological Characteristics

The SCSK is located in the north of the East China Sea, which represents the western arm of the Pacific Ocean. The rocks of the region are igneous, metamorphic and sedimentary types, of which aged Precambrian gneiss and schist are the oldest [19]. Tertiary Pleistocene strata are widespread, and large deltas and marine terrace deposits can be found along the coasts. Several small, NNE-trending basins have been found, with a large area in the intermediate basement [20]. A previous study noted that a sequence of Late Cretaceous consists of fluviolacustrine, which is affected by volcanic activity [21].

Seismic reflections from the sequence between the Late Cretaceous and Late Eocene/Early Oligocene are low to very high in amplitude, showing the various depositional processes. The volcanic intrusion is thought to date back to the Early Miocene, and the flow was covered by silt and siltstone until the Late Miocene [20,21]. The sedimentary layer below the basalt eruption is mainly composed of unconsolidated inter-bedding sandstones, claystones and occasionally siltstones, attributed to the subsidence of the Jeju Basin during the Late Eocene to Early Oligocene [22]. A recent study focused on the potential for CO_2 storage using the basalt flow's sealing ability above the sedimentary layer [16]. In 2018, the Korea Institute of Geoscience and Mineral Resources (KIGAM) carried out a two-dimensional seismic survey on the Jeju Basin, north of the East China Shelf Basin, to assess its potential for the storage of CO_2 (Figure 1a). A total of nine multichannel seismic lines were surveyed, covering an area of 130 square kilometers (Figure 1b).

The two formations of interest in this study were volcanic mounds and the lower depositional layer, which can be cap rocks and storage for CO_2. The results of the well-log analysis showed that the petro-physical properties of good permeability and porosity are important for potential CO_2 storage [23]. The top volcanic mounds were identified by anomaly in the well-log analysis for sonic and porosity properties (Figure 2). A vertical-depth error can occur and/or be accumulated in the seismic data that normally are recorded and displayed in the time-domain vertical axis.

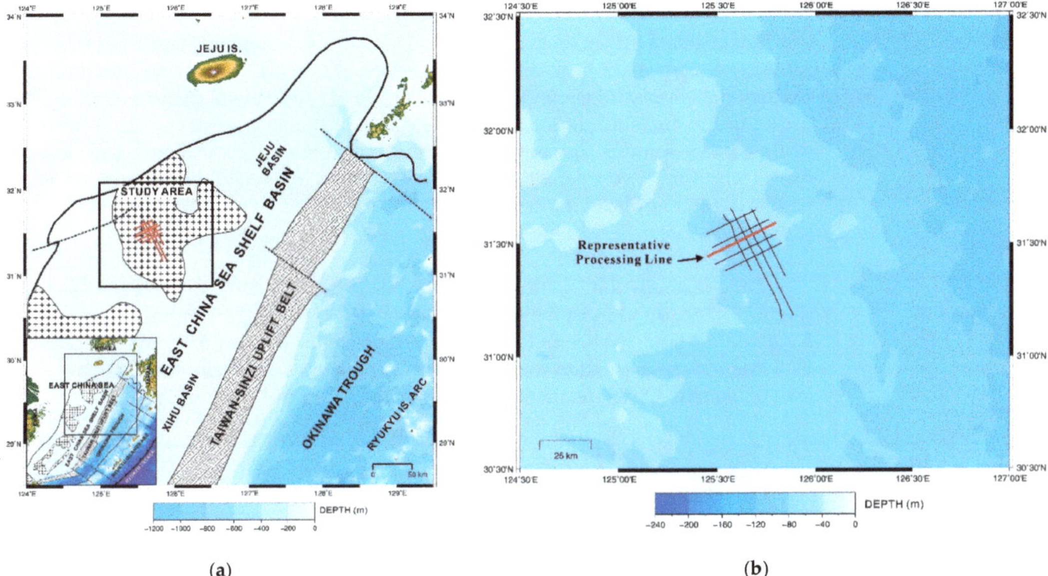

(a) (b)

Figure 1. (**a**) Survey area map and (**b**) seismic acquisition lines (modified from [24]). The bold red line is a representative seismic processing line.

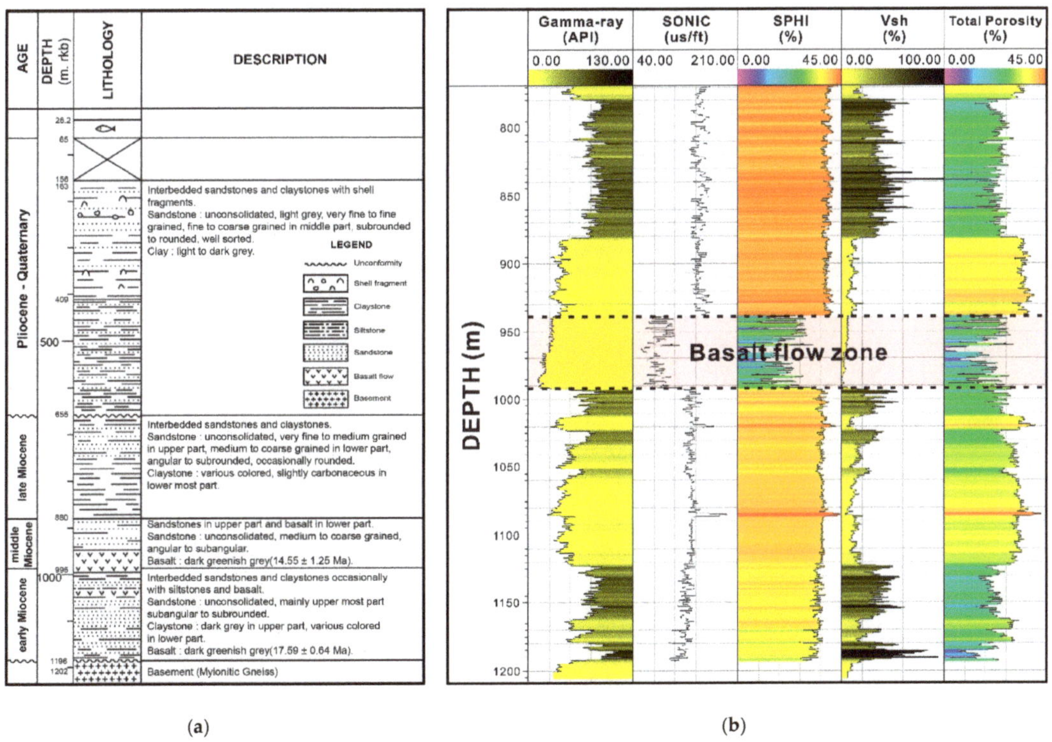

(a) (b)

Figure 2. (**a**) Geological summary of previous well-log of an adjacent study area (modified from [23]) and (**b**) well-log analysis of (**a**) with storage potential zone. SPHI: sonic-derived porosity. Vsh: shale volume.

3. Results of Seismic Data Processing

Seismic data processing consists of time- and depth-domain stages; the latter includes MVA. The purpose of time-domain processing is to produce suitable subsurface images to carry out successful geological interpretation of CO_2 storage feasibility. Depth-domain processing aims to accurately refine the depth of the subsurface, which is useful in the drilling process and in the computation of storage volume. The following sections explain how each of the two processing steps is executed, including the input parameters and results of filter application.

3.1. Time-Domain Processing

Time-domain data processing was performed on offshore seismic survey lines (Figure 1b) acquired from the study area (Figure 1a) using the Tamhae2 research vessel of KIGAM. The bold red line in Figure 1b is a representative seismic line, with approximately 37.5 line-km. Seismic data were acquired using a 1.2 km (96 channels) streamer and the air-gun source. Other detailed acquisition parameters are listed in Table 1.

Table 1. Seismic data acquisition parameters.

Parameter	Description	Value
Streamer	Streamer length (m)	1200
	No. of channels	96
	Group interval (m)	12.5
	Offset ranges (m)	125~1325
	Streamer depth (m)	7 ± 1
Recording	Recording length (s)	5
	Sampling rate (ms)	1
Source	Source type	Bolt long-life air guns
	Source volume (cu. in.)	1254
	Source depth (m)	5
	Source interval (m)	25

To confirm the effectiveness of the time-domain data processing techniques (Figure 3a) applied in this paper, the results before and after the signal processing were compared (Figures 4 and 5). Figures 4a and 5a show the shot gather and the stack section before the signal processing, and there is a multiple (refer to the arrows) at around 1.05 s. We can confirm from the results that the multiple was effectively attenuated (Figures 4b and 5b) after the signal processing was applied. In addition, the amplitude spectra before (black line in Figure 4c) and after (red line in Figure 4c) the signal processing were compared to confirm the effectiveness of time-domain data processing. From Figure 4c, we can confirm that the low-frequency signals were enhanced and the notches at higher frequencies were removed after the signal-processing techniques were applied. Therefore, through the application of time-domain data processing, the multiples and noise inherent in the field data can be effectively attenuated. In addition, from the stack section (Figure 5b), after applying the time-domain processing, the connectivity of the horizontal layers in the region of interest (0.9 to 1.2 s) greatly improved, and detailed subsurface structures could be imaged.

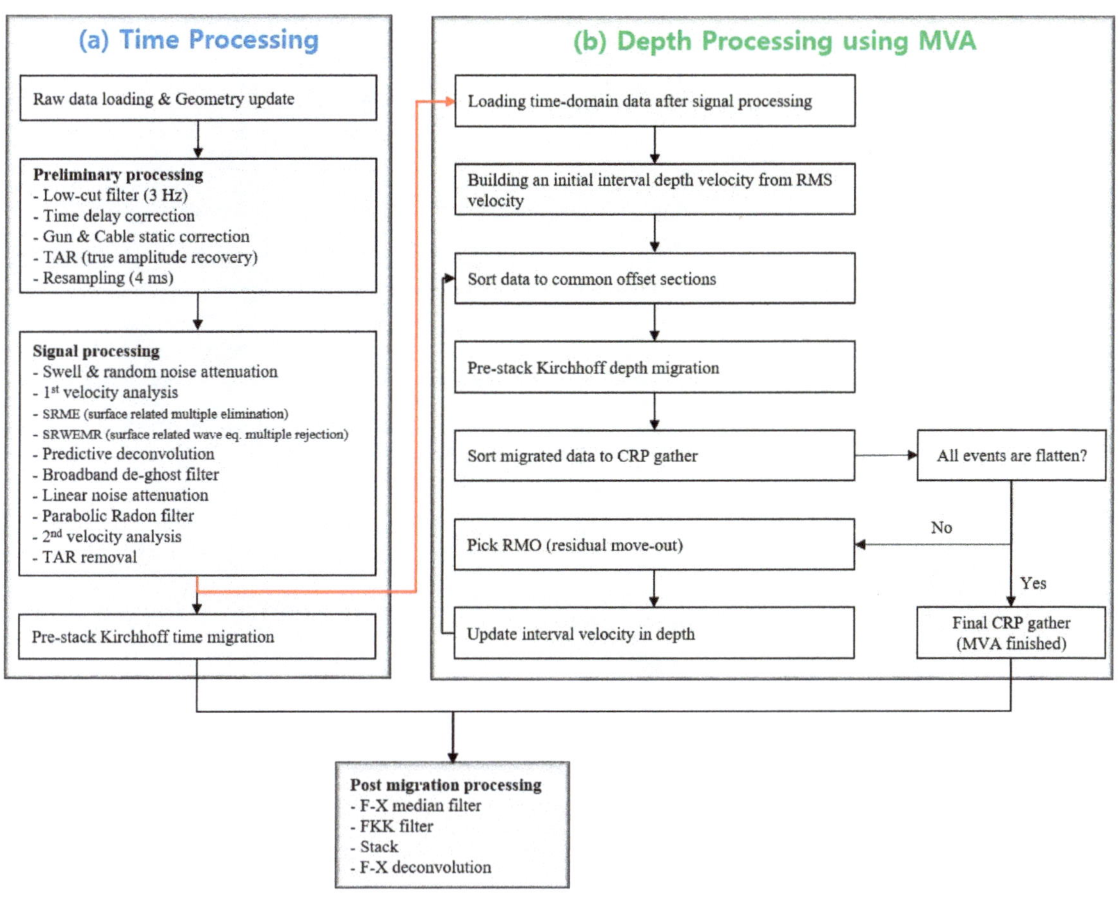

Figure 3. Seismic data processing workflows in the (**a**) time and (**b**) depth domains.

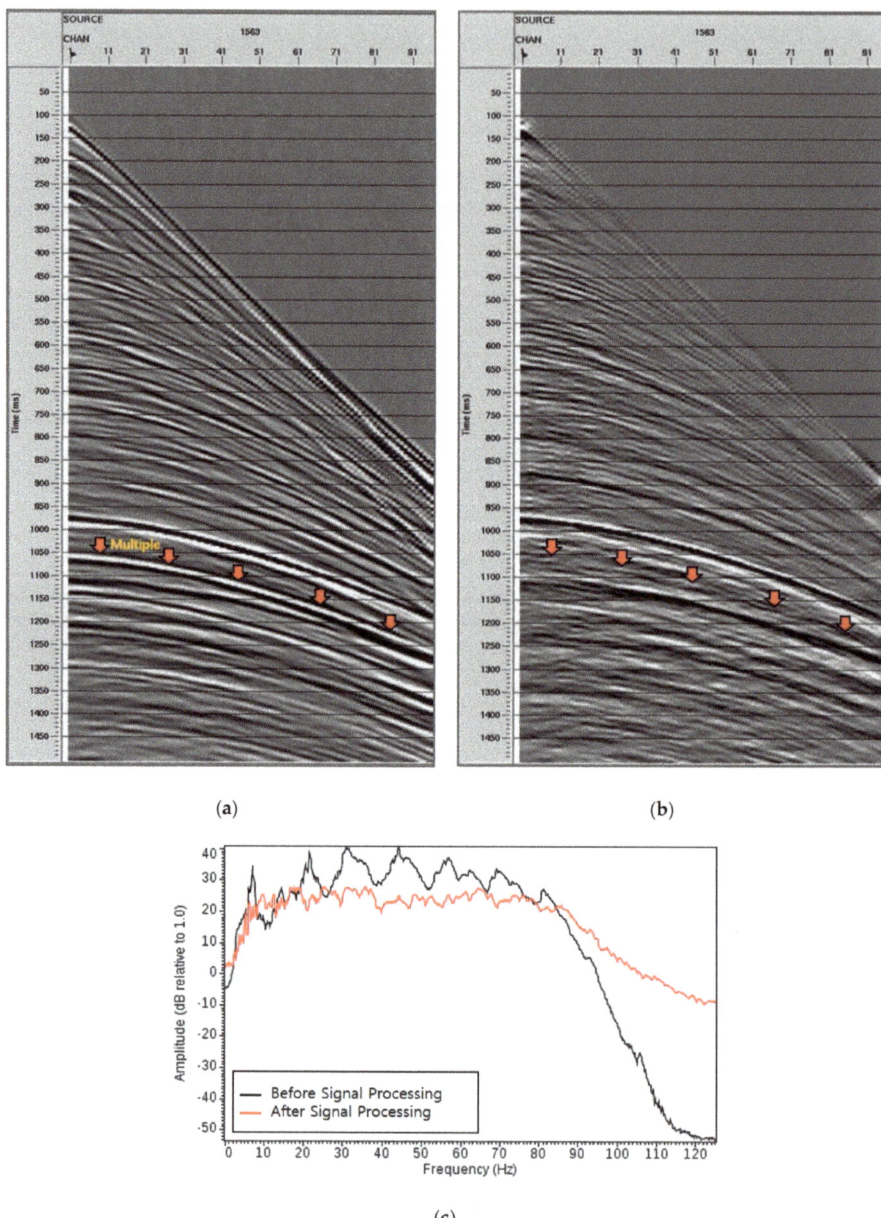

Figure 4. Shot gathers (**a**) before and (**b**) after the signal processing. (**c**) Amplitude spectra before (black line) and after (red line) the signal processing.

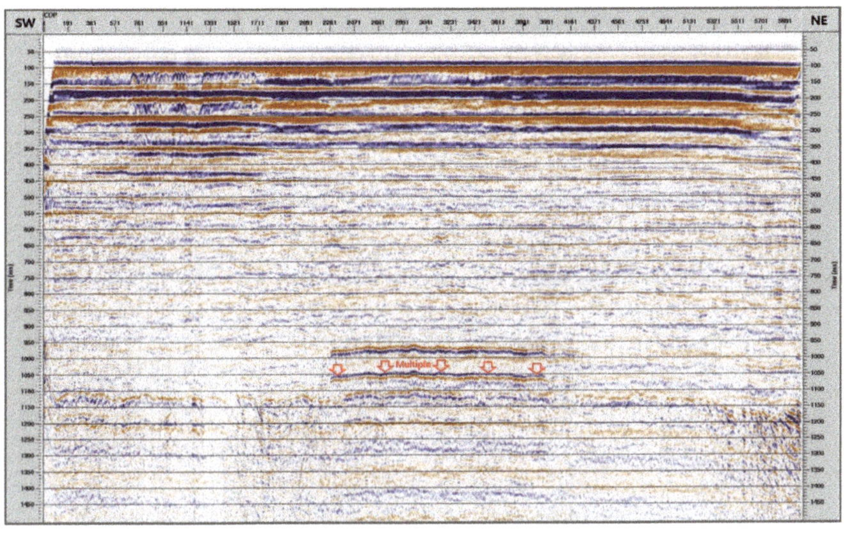

(**a**)

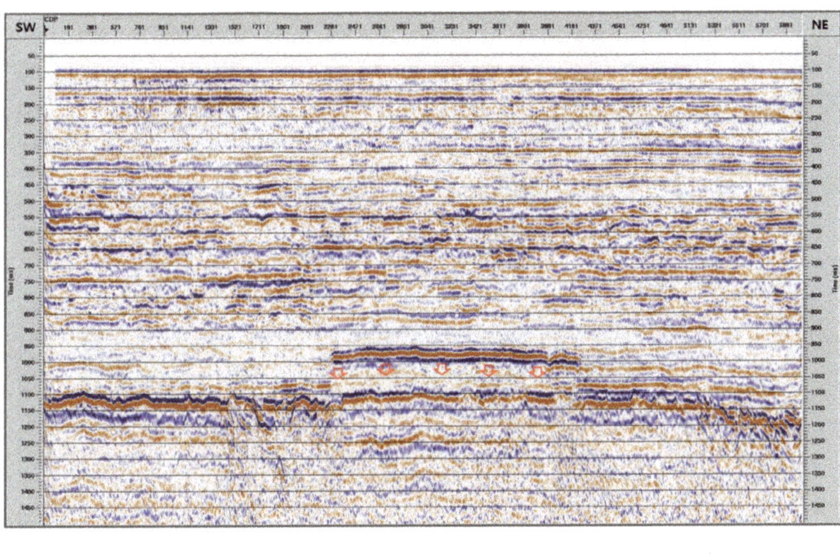

(**b**)

Figure 5. (**a**) Stack section after applying only the preliminary processing in time-processing workflow. (**b**) Stack section after applying the entire time-processing workflow (Figure 3a).

3.2. Depth-Domain Processing Using the MVA Technique

Depth-domain data processing flattened the CRP gathers, which proves that the updated velocity is correct. The MVA technique is a data-processing method that can obtain a reasonable velocity estimation using tomography or a wave equation for the complex subsurface structures. Then, this velocity is used to generate a subsurface image in the depth domain. In this study, a depth-domain data-processing workflow (Figure 3b) using a tomography-based MVA technique [18] was applied. In order to obtain high-quality depth-domain results, e.g., seismic sections and velocities, the input data were used with data (Figure 4b) whose signal-to-noise ratio was improved after applying the signal processing in the time domain. The depth-domain interval velocity, converted from the time-domain RMS velocity by the Dix equation [25], was used as the initial velocity model. Next, the loaded time-domain data were sorted by the common offset data. After calculating the travel time [26] using the eikonal solver, we performed Kirchhoff pre-stack depth migration (PSDM) [27]. The migration results were sorted as CRP gathers, and the residual move-out (RMO) technique [28] was applied. Finally, ray-based tomographic velocity analysis [18] was applied to obtain the updated depth-domain velocity.

In this study, depth-domain processing using the MVA technique (Figure 3b) was repeated three times. By comparing the depth-domain initial velocity (Figure 6a) converted from the RMS velocity using time-depth conversion and the final velocity (Figure 6b) obtained after MVA at the third iteration, we were able to confirm that the overall subsurface velocity structures after applying MVA were greatly improved. In particular, it can be seen from the velocity differences in Figure 6c that the velocity was significantly updated in the region of interest (0.9 to 1.2 km), which corresponds to the depth of basalt flow.

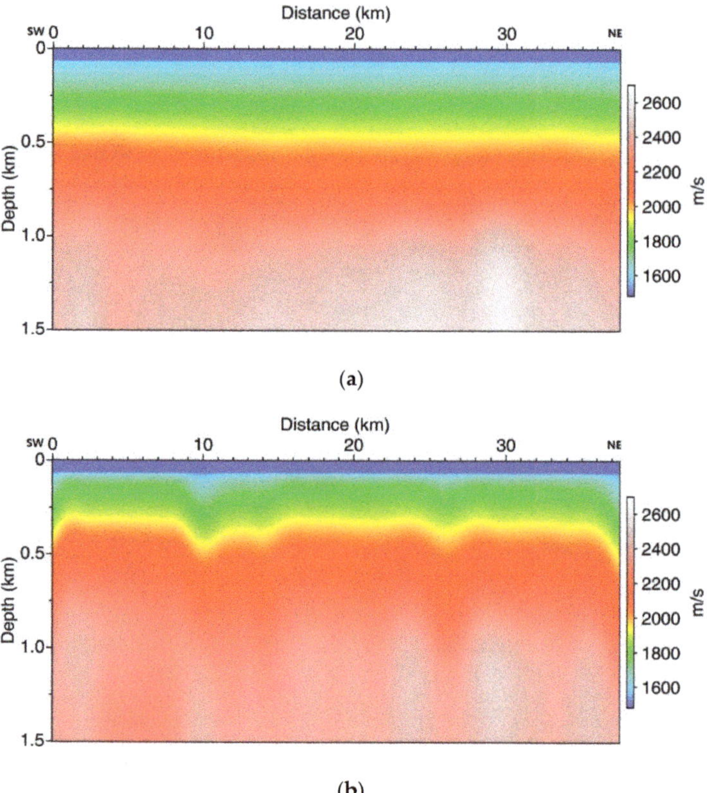

(a)

(b)

Figure 6. *Cont.*

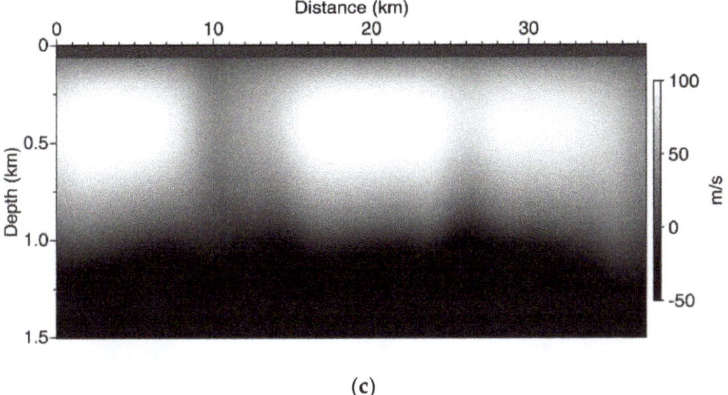

(c)

Figure 6. (**a**) Initial depth-domain interval velocity for MVA obtained by converting the time-domain RMS velocity. (**b**) Final updated depth-domain interval velocity after MVA at the 3rd iteration. (**c**) Difference between (**a**) and (**b**).

In general, the CRP gather is analyzed to verify whether the velocity is correct. CRP gathers are signals reflected from the same reflection point, so if the velocity is correct, the CRP gather should be flat [28]. Therefore, we compared CRP gathers before (Figure 7a) and after (Figure 7b) MVA to confirm that the final depth-domain velocity was reasonable. From Figure 7, we can observe that the CRP gather became flatter after the MVA was applied. Since the CRP gather was generated depending on the velocity, the depth of the reflection events in the CRP gather was corrected using the updated velocity after applying the MVA. In addition, in order to quantitatively confirm the accuracy of the updated velocity, the velocity error curve according to the number of iterations is shown in Figure 8. In general, the iteration number is determined by considering the trade-off between accuracy and computational cost. It can be seen that the velocity error was reduced to about 3.8% in the third iteration. From the error curve, we judged that the velocity error sufficiently converged in the third iteration, and thus the MVA process was stopped. Therefore, it was verified that the final depth-domain velocity obtained after applying the MVA was reasonable.

Finally, to confirm the effectiveness of the depth-domain data processing, we compared the stack sections before (Figure 9a) and after (Figure 9b) the MVA was applied. The red and green lines represent the top of the basalt before and after MVA was applied, respectively. In Figure 9b, the difference in depth between the two basalt tops ranges from 32.5 to 60 m. The 3D depth-domain stack sections after applying MVA are also shown in Figure 10. From the 3D results, the horizons between the in-line and the cross-line agree well with each other, and the spatial depth variance of the basalt top before and after the MVA application can be observed. Therefore, we confirmed that not only the velocity but also the subsurface structure was corrected after applying the depth-domain data processing using the MVA technique.

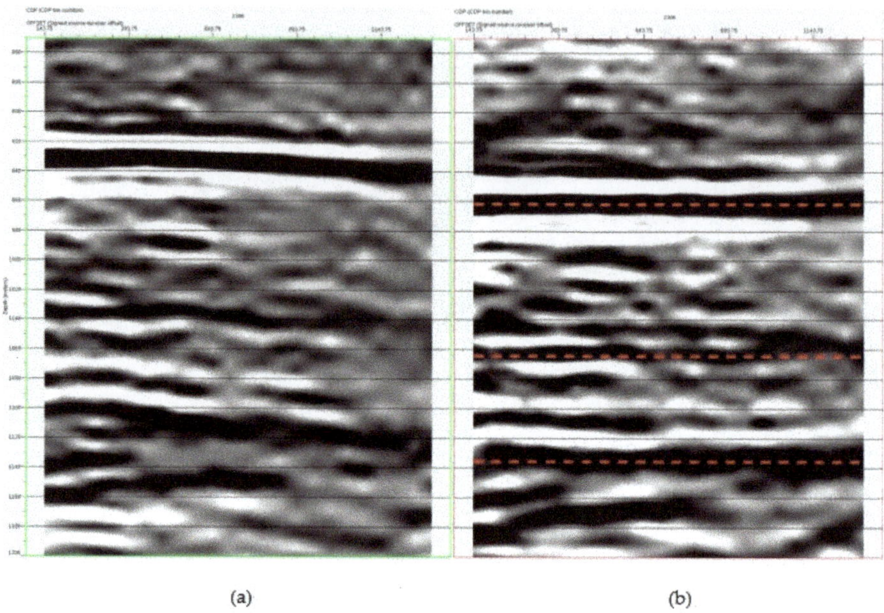

(a)   (b)

Figure 7. (**a**) CRP gather before MVA and (**b**) CRP gather after MVA at the third iteration.

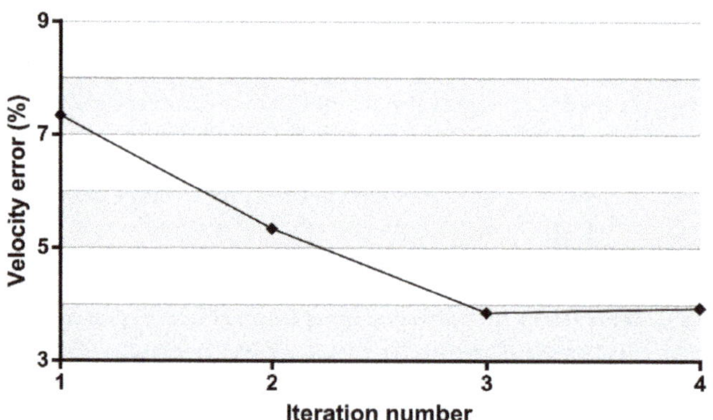

Figure 8. Velocity error curve at a depth of 960 m in the 2705th CRP gather.

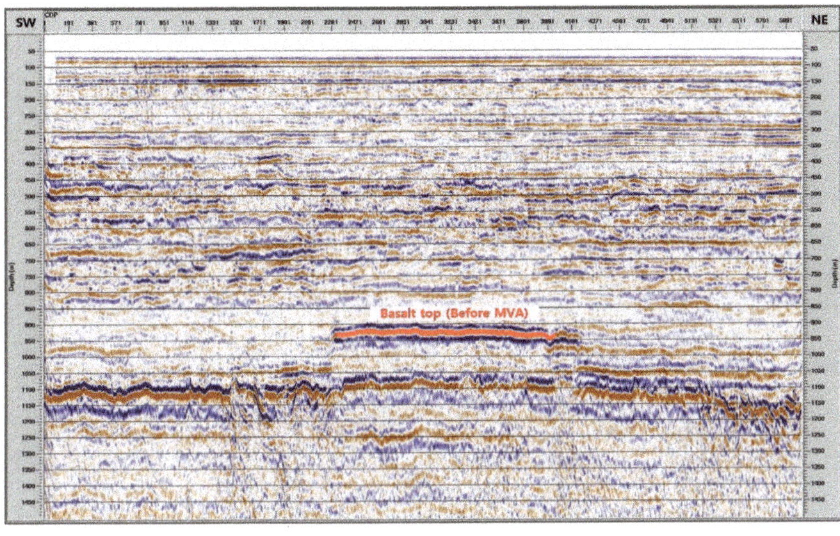

(**a**)

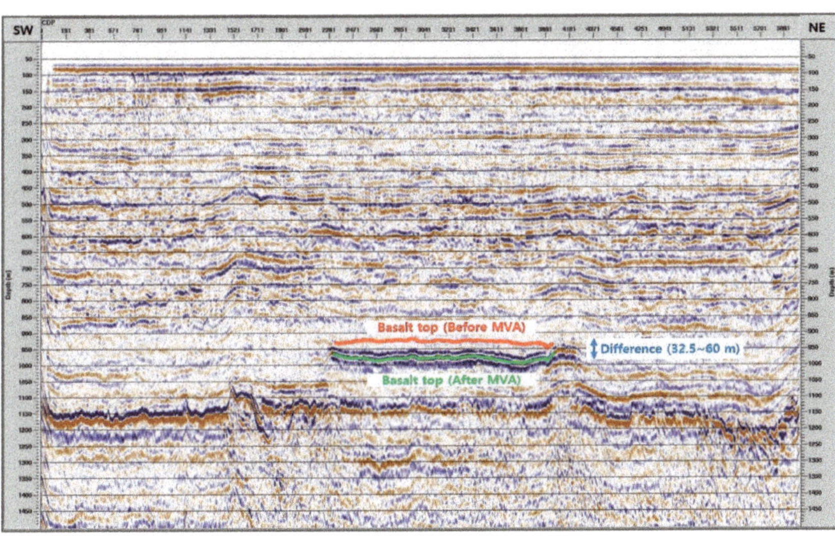

(**b**)

Figure 9. (**a**) Depth-domain stack section obtained by converting the time-domain result (Figure 5b). (**b**) Depth-domain stack section after applying MVA at the 3rd iteration. The red and green lines represent the top of the basalt before and after applying MVA, respectively. The difference in depth between the two basalt tops ranges from 32.5 to 60 m.

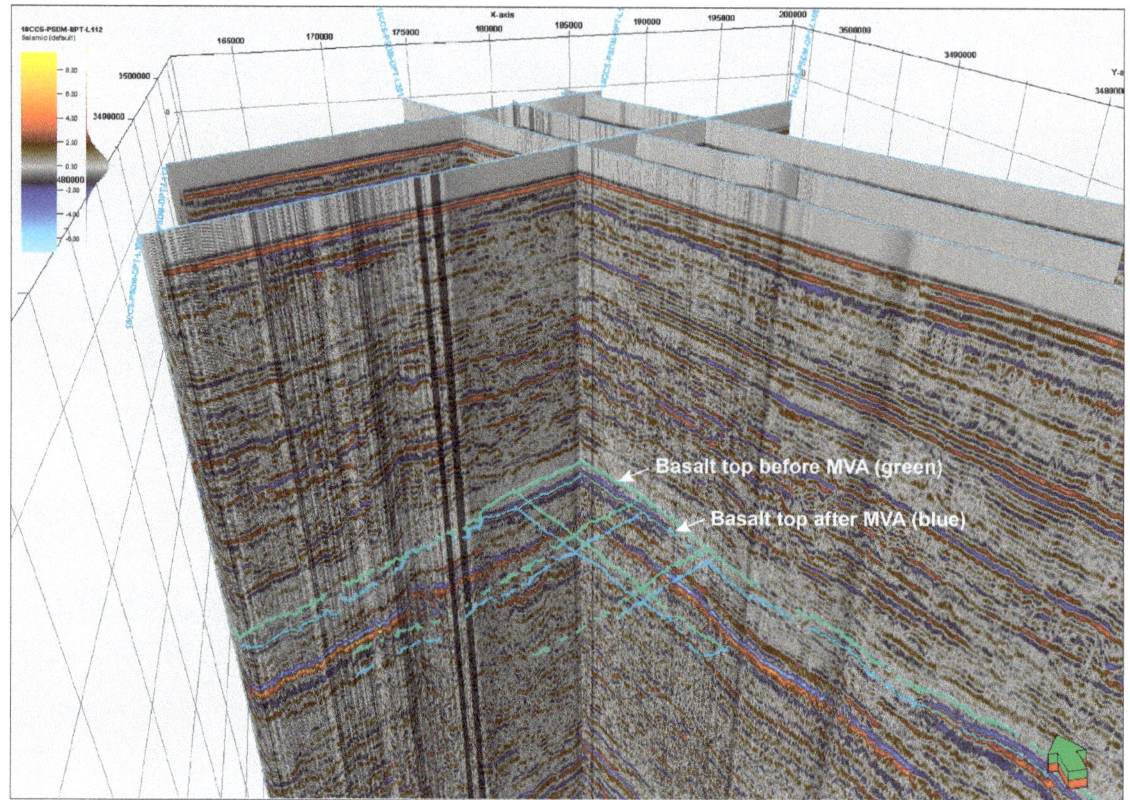

Figure 10. 3D depth-domain stack sections after applying MVA at the 3rd iteration. The green and blue lines represent the top of the basalt before and after applying MVA, respectively.

4. Seismic Interpretation

In order to compare the results obtained before and after the depth processing using the MVA technique was applied, we interpreted a reflector, which was strong amplitude (Figure 11). We picked the same reflector from the results obtained both before and after depth processing was applied, and the reflectors are represented by contour maps in Figure 11a,b. We confirmed the boundary using the seismic attribute of RMS analysis, indicating an amplitude anomaly (Figure 11c,d). The RMS seismic attribute is a post-stack analysis technique and computes the square root of the sum of squared amplitudes within the specified window used.

The strong reflector from time processing (upper image in Figure 12a) vertically ranged from 910 to 980 m, while the reflector from both time and depth processing (lower image in Figure 12a) was mainly positioned below, ranging from approximately 970 to 1020 m. The reflector was interpreted as the top of the volcanic mound. The volcanic body showed high acoustic impedance compared with the surrounding clastic rocks, and products derived from igneous rocks could be easily detected in sedimentary basins. The reflector constrained to the RMS seismic attribute showed a circle-shaped geometry and the volcanic body was located high in the center, indicating a volcanic mound (Figure 11a,c). Compared with mound shape, in both time and depth processing (Figure 11b,d), lateral changes were presented in greater detail, although the mound subjected to only time processing (Figure 11a,c) was smoother. As a result of the interpretation, the results (Figure 12a) from both time and depth processing using MVA show that the volcanic mound vertically shifted to a deeper position. The difference (Figure 12b) in the mound before and after depth processing was applied was varied, ranging from 32.5 to 60 m. We suggest that geological geometry or objects derived from both time and depth processing using MVA represent vertical differences and show lateral variations (Figure 12b).

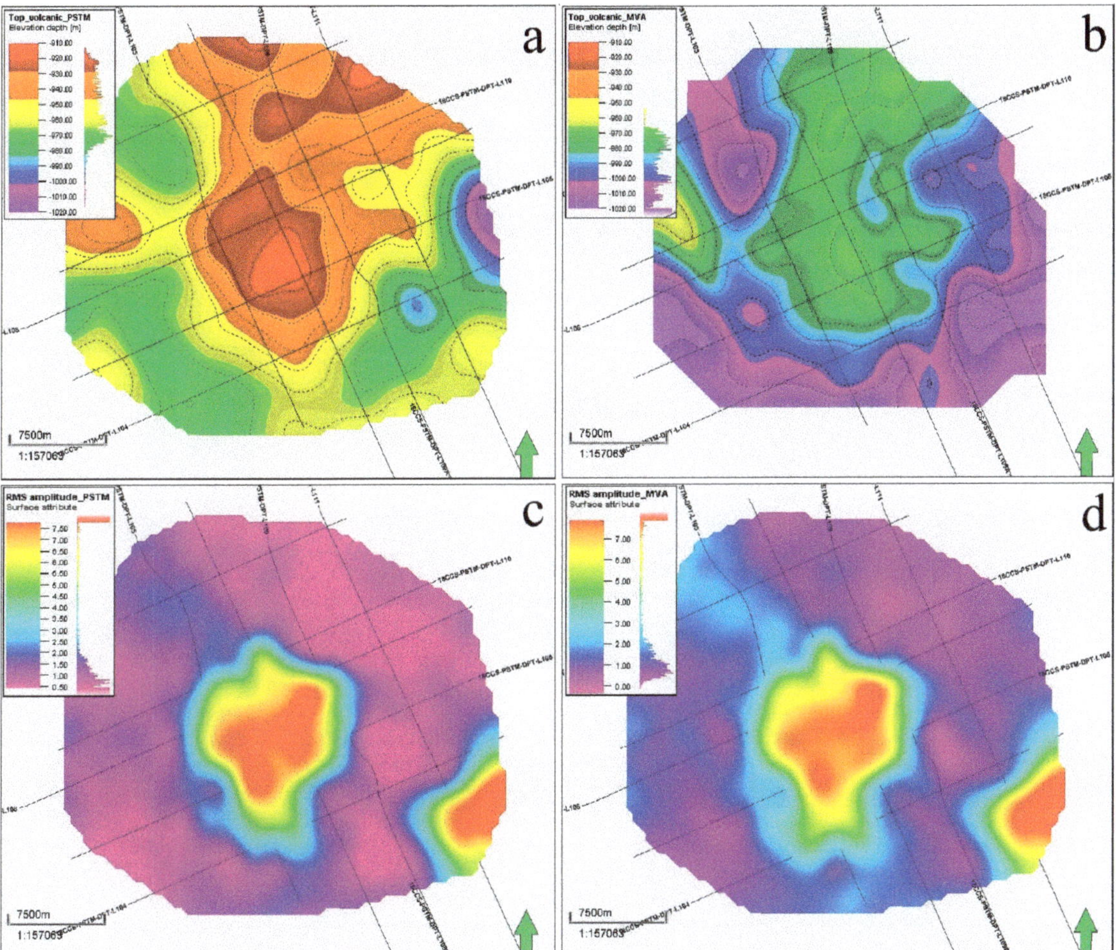

Figure 11. Contour maps of the strong reflector and RMS attributes. Geologically interpreted tops of the volcanic mounds (**a**) before and (**b**) after depth processing. RMS amplitude attributes (**c**) before and (**d**) after depth processing showing volcanic mounds' anomalous region.

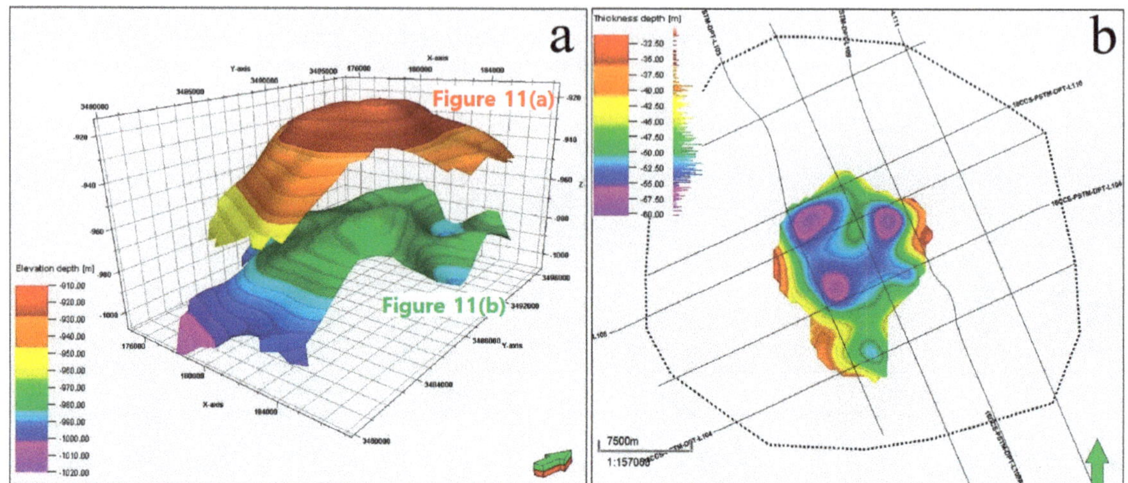

Figure 12. (**a**) Overlapping tops of the volcanic mounds before (Figure 11a) and after (Figure 11b) the depth processing. (**b**) The contour map represents vertical differences between Figure 11a,b.

5. Discussion

The direct subsurface information of a volcanic area can be acquired from the results of analysis by drilling a core and well-logging. In the research area, we unfortunately did not have adequate drilling or logging data. From the seismic survey data, we inferred a basalt flow with strong amplitude contrast but could not determine the precise depth of the target structure. Various geophysical methods pose challenges to acquiring the basalt depth from indirect survey data. We focused on improving the velocity by using composite seismic data-processing modules. MVA yielded accurate velocity information, and it is one of the most innovative techniques among the tomography-based methods. Another option for successful velocity calculation is the full waveform inversion method [29,30], which is based on wave equation extrapolation. The selection of a velocity-calculation method might be dependent on the available seismic data and geological interpretation.

The present seismic data-processing flow showed increased horizontal continuity with regard to expansion phenomena, which leads to a reduced elevation range but a more complicated shape. If more dense, gridded 3D seismic exploration data can be obtained and processed, and more specific basalt eruptions can be observed in seismic images. Even in the case of a 3D survey, the large aspect of the horizontal change will be similar and the vertical depth will be similar, as demonstrated by the present results.

One challenge that remains unresolved is the tracking of changes in the anisotropic medium. Although the basalt flow shows less anisotropy, it is necessary to accurately consider the anisotropy of the cap rocks and images because it has migrated from the eruption location. To consider the anisotropy, it is recommended to record the seismic reflection signal using a streamer for as long as possible. With large offset data, anisotropic MVA [31,32] will be possible, enabling the boundary of the basalt flow to be defined more accurately.

6. Conclusions

The resolution of the basalt flow structure was improved by the use of optimized time- and depth-domain data processing of offshore seismic data regarding the use of cap rocks for CO_2 storage. Time processing yielded a subsurface image section with an improved signal-to-noise ratio by attenuating multiples and noise. Depth processing refined the velocity distribution so that the depth of the basalt flow was corrected by the lateral and vertical axis. Depth-domain processing used the MVA technique to update the velocity by the iteration of the travel-time calculation. Flattened CRP gathers qualitatively proved that the results of the processing approached the exact depth of the basalt flow. To quantify the accuracy of MVA, we confirmed that the velocity error stably decreased until third iteration. The slice map showed the aspect of basalt intrusion in detail and suggested that the cap rock was located at a greater depth of approximately 32.5–60 m compared with when only time-domain processing was applied. Comprehensive seismic data processing including both the time and depth domains is required when a relatively deep subsurface structure resulting from volcanic activity is to be imaged.

Author Contributions: Conceptualization, S.C. and M.K.; methodology, W.S. and S.C.; software, W.S. and C.L.; validation, W.S., S.C. and M.K.; investigation, S.C.; data curation, S.C. and M.K.; writing—original draft preparation, W.S., S.C. and C.L.; writing—review and editing, W.S., S.C. and M.K. All authors have read and agreed to the published version of the manuscript.

Funding: This research was funded by the Korea Institute of Geoscience and Mineral Resources (KIGAM), grant numbers 21-3413-1 (Technology development for storage efficiency improvement and safety assessment of CO_2 geological storage) and 21-3312-1 (Development of the integrated geophysical survey and real-scale data processing technologies for 3D high-resolution imaging of the marine subsurface).

Institutional Review Board Statement: Not applicable.

Informed Consent Statement: Not applicable.

Data Availability Statement: Not applicable.

Acknowledgments: The software Generic Mapping Tools was used to plot the map of the study area [33].

Conflicts of Interest: The authors declare no conflict of interest.

References

1. US Government. How Much of the Earth is Volcanic? Available online: https://www.usgs.gov/faqs/how-much-earth-volcanic?qt-news_science_products=0#qt-news_science_products (accessed on 18 March 2021).
2. Blaikie, T.N.; Ailleres, L.; Betts, P.G.; Cas, R.A.F. Interpreting subsurface volcanic structures using geologically constrained 3-D gravity inversions: Examples of maar-diatremes, Newer Volcanics Province, Southeastern Australia. *J. Geophys. Res. Solid Earth* **2014**, *119*, 3857–3878. [CrossRef]
3. Aiello, G.; Angelino, A.; D'Argenio, B.; Marsella, E.; Pelosi, N.; Ruggieri, S.; Siniscalchi, A. Buried volcanic structures in the Gulf of Naples (Southern Tyrrhenian Sea, Italy) resulting from high resolution magnetic survey and seismic profiling. *Ann. Geophys.* **2005**, *48*, 883–897.
4. Tsuru, T.; No, T.; Fujie, G. Geophysical imaging of subsurface structures in volcanic area by seismic attenuation profiling. *Earth Planets Space* **2017**, *69*, 1–12. [CrossRef]
5. Klamer, S.; Klamer, O. Identification of paleo-volcanic rocks on seismic data. In *Updates in Volcanology—A Comprehensive Approach to Volcanological Problems*, 1st ed.; Stoppa, F., Ed.; InTech: Rijeka, Croatia, 2012.
6. Bruno, P.P.G.; Cippitelli, G.; Rapolla, A. Seismic study of the Mesozoic carbonate basement around Mt. Somma-Vesuvius, Italy. *J. Volcanol. Geotherm. Res.* **1998**, *84*, 311–322. [CrossRef]
7. Lomax, A.; Zollo, A.; Capuano, P.; Virieux, J. Precise, absolute earthquake location under Somma-Vesuvius volcano using a new three-dimensional velocity model. *Geophys. J. Int.* **2001**, *146*, 313–331. [CrossRef]
8. Hooft, E.E.E.; Heath, B.A.; Toomey, D.R.; Paulatto, M.; Papazachos, C.B.; Nomikou, P.; Morgan, J.V.; Warner, M.R. Seismic imaging of Santorini: Subsurface constraints on caldera collapse and present-day magma recharge. *Earth Planet. Sci. Lett.* **2019**, *514*, 48–61. [CrossRef]
9. Tartarello, M.C.; Palisant, A.; Bigi, S.; Beaubien, S.E.; Graziani, S.; Lombardi, S.; Ruggiero, L.; De Angelis, D.; Sacco, P.; Maggio, E. Preliminary results of geological characterization and geochemical monitoring of Sulcis Basin (Sardinia), as a potential CCS site. *Ener. Proc.* **2017**, *125*, 549–555. [CrossRef]
10. McGrail, B.P.; Schaef, H.T.; Ho, A.M.; Chien, Y.; Dooley, J.J.; Davidson, C.L. Potential for carbon dioxide sequestration in flood basalts. *J. Geophys. Res. Space Phys.* **2006**, *111*, B12201. [CrossRef]
11. Busch, A.; Amann, A.; Bertier, P.; Waschbusch, M.; Krooss, B.M. The significance of caprock sealing integrity for CO_2 storage. In Proceedings of the SPE International Conference on CO_2 Capture, Storage and Utilization, New Orleans, LA, USA, 10–12 November 2010.

12. Matter, J.M.; Stute, M.; Snaebjornsdottir, S.O.; Oelkers, E.H.; Gislason, S.R.; Aradottir, E.S.; Sigfusson, B.; Gunnarsson, I.; Sigurdardottir, H.; Gunnlaugsson, E.; et al. Rapid carbon mineralization for permanent disposal of anthropogenic carbon dioxide emissions. *Science* **2016**, *352*, 1312–1314. [CrossRef]

13. Jayne, R.S.; Wu, H.; Pollyea, R.M. Geologic CO_2 sequestration and permeability uncertainty in a highly heterogeneous reservoir. *Int. J. Green. Gas. Cont.* **2019**, *83*, 128–139. [CrossRef]

14. Kang, M.; Kim, J.H.; Kim, K.; Cheong, S.; Shinn, Y.J. Assessment of CO_2 storage capacity for basalt caprock-sandstone reservoir system in the northern East China Sea. In Proceedings of the EGU General Assembly Conference Abstract, Vienna, Austria, 8–13 April 2018; p. 5772.

15. Huh, D.; Park, Y.; Yoo, D.; Hwang, S. CO_2 geological storage potential in Korea. *Ener. Proc.* **2011**, *4*, 4881–4888. [CrossRef]

16. Litt, T.; Krastel, S.; Sturm, M.; Kipfer, R.; Orcen, S.; Heumann, G.; Franz, S.O.; Ulgen, U.B.; Niessen, F. 'PALEOVAN', International Continental Scientific Drilling Program (ICDP): Site survey results and perspectives. *Quat. Sci. Rev.* **2009**, *28*, 1555–1567. [CrossRef]

17. Stork, C.; Clayton, R.W. Linear aspects of tomographic velocity analysis. *Geophysics* **1991**, *56*, 483–495. [CrossRef]

18. Stork, C. Reflection tomography in the postmigrated domain. *Geophysics* **1992**, *57*, 680–692. [CrossRef]

19. Niino, H.; Emery, K.O. Sediments of shallow portions of East China Sea and South China Sea. *Geol. Soc. Am. Bull.* **1961**, *72*, 731–762. [CrossRef]

20. Cukur, D.; Horozal, S.; Kim, D.C.; Han, H.C. Seismic stratigraphy and structural analysis of the northern East China Sea Shelf Basin interpreted from multi-channel seismic reflection data and cross-section restoration. *Mar. Pet. Geol.* **2011**, *28*, 1003–1022. [CrossRef]

21. Lee, G.H.; Kim, B.; Shin, K.S.; Sunwoo, D. Geologic evolution and aspects of the petroleum geology of the northern East China Sea shelf basin. *AAPG Bull.* **2006**, *90*, 237–260. [CrossRef]

22. Lee, C.; Shinn, Y.J.; Han, H.-C. Structural evolution of two-stage rifting in the northern East China Sea Shelf Basin. *Geol. J.* **2019**, *54*, 2229–2240. [CrossRef]

23. Korea Institute of Geoscience and Mineral Resources. *Hydrocarbon Potential of the Northern East China Shelf Basin, I*; KIGAM Report; KIGAM: Daejeon, Korea, 1997; (written in Korean with English abstract).

24. Shin, S.Y.; Kang, M.; Shinn, Y.J.; Cheong, S. Assessment of CO_2 geological storage capacity for basalt flow structure around PZ-1 exploration well in the Southern Continental Shelf of Korea. *Econ. Environ. Geol.* **2020**, *53*, 33–43. (written in Korean with English abstract)

25. Dix, C.H. Seismic velocities from surface measurements. *Geophysics* **1955**, *20*, 68–86. [CrossRef]

26. Vidale, J. Finite-difference calculation of traveltimes in three dimensions. *Geophysics* **1990**, *55*, 521–526. [CrossRef]

27. Gray, S.H.; May, W.P. Kirchhoff migration using eikonal equation traveltimes. *Geophysics* **1994**, *59*, 810–817. [CrossRef]

28. Yilmaz, O. *Seismic Data Analysis: Processing, Inversion, and Interpretation of Seismic Data*, 2nd ed.; SEG: Tulsa, OK, USA, 2001.

29. Virieux, J.; Operto, S. An overview of full-waveform inversion in exploration geophysics. *Geophysics* **2009**, *74*, WCC127–WCC152. [CrossRef]

30. Son, W.; Pyun, S.; Shin, C.; Kim, H. Laplace-domain wave-equation modeling and full waveform inversion in 3D isotropic elastic media. *J. Appl. Geophys.* **2014**, *105*, 120–132. [CrossRef]

31. Zhou, H.; Pham, D.; Gray, S. Tomographic velocity analysis in strongly anisotropic TTI media. In Proceedings of the 74th Annual International Meeting of the SEG, Expanded Abstracts, Denver, CO, USA, 10–15 October 2004; pp. 2347–2350.

32. Li, Y.; Biondi, B. Migration velocity analysis for anisotropic models. In Proceedings of the 81st Annual International Meeting of the SEG, Expanded Abstracts, San Antonio, TX, USA, 18–23 September 2011; pp. 201–206.

33. Wessel, P.; Smith, W.H.F.; Scharroo, R.; Luis, J.; Wobbe, F. Generic mapping tools: Improved version released. *EOS Trans. AGU* **2013**, *94*, 409–410. [CrossRef]

applied
sciences

Article

Implementation of Robust Satellite Techniques for Volcanoes on ASTER Data under the Google Earth Engine Platform

Nicola Genzano [1,*], Francesco Marchese [2], Marco Neri [3], Nicola Pergola [2] and Valerio Tramutoli [1]

[1] Scuola di Ingegneria, Università degli Studi della Basilicata, 85100 Potenza, Italy; valerio.tramutoli@unibas.it
[2] Istituto di Metodologie per l'Analisi Ambientale (IMAA), Consiglio Nazionale delle Ricerche (CNR), 85050 Tito Scalo, Italy; francesco.marchese@imaa.cnr.it (F.M.); nicola.pergola@imaa.cnr.it (N.P.)
[3] Istituto Nazionale di Geofisica e Vulcanologia, Osservatorio Etneo, 95125 Catania, Italy; marco.neri@ingv.it
* Correspondence: nicola.genzano@unibas.it; Tel.: +39-0971-205-047

Featured Application: Volcano activity mapping and monitoring.

Abstract: The RST (Robust Satellite Techniques) approach is a multi-temporal scheme of satellite data analysis widely used to investigate and monitor thermal volcanic activity from space through high temporal resolution data from sensors such as the Moderate Resolution Imaging Spectroradiometer (MODIS), and the Spinning Enhanced Visible and Infrared Imager (SEVIRI). In this work, we present the results of the preliminary RST algorithm implementation to thermal infrared (TIR) data, at 90 m spatial resolution, from the Advanced Spaceborne Thermal Emission and Reflection Radiometer (ASTER). Results achieved under the Google Earth Engine (GEE) environment, by analyzing 20 years of satellite observations over three active volcanoes (i.e., Etna, Shishaldin and Shinmoedake) located in different geographic areas, show that the RST-based system, hereafter named RASTer, detected a higher (around 25% more) number of thermal anomalies than the well-established ASTER Volcano Archive (AVA). Despite the availability of a less populated dataset than other sensors, the RST implementation on ASTER data guarantees an efficient identification and mapping of volcanic thermal features even of a low intensity level. To improve the temporal continuity of the active volcanoes monitoring, the possibility of exploiting RASTer is here addressed, in the perspective of an operational multi-satellite observing system. The latter could include mid-high spatial resolution satellite data (e.g., Sentinel-2/MSI, Landsat-8/OLI), as well as those at higher-temporal (lower-spatial) resolution (e.g., EOS/MODIS, Suomi-NPP/VIIRS, Sentinel-3/SLSTR), for which RASTer could provide useful algorithm's validation and training dataset.

Keywords: volcanoes; ASTER; Robust Satellite Techniques; Google Earth Engine

Citation: Genzano, N.; Marchese, F.; Neri, M.; Pergola, N.; Tramutoli, V. Implementation of Robust Satellite Techniques for Volcanoes on ASTER Data under the Google Earth Engine Platform. *Appl. Sci.* **2021**, *11*, 4201. https://doi.org/10.3390/app11094201

Academic Editor: Alessandro Bonforte

Received: 2 April 2021
Accepted: 1 May 2021
Published: 5 May 2021

1. Introduction

Several studies have shown the relevant contribution of satellite systems for investigating and monitoring active volcanoes, also in areas where ground-based devices are fully operational (e.g., [1–4]). Those systems are capable of providing information about changes of thermal volcanic activity leading to eruptions and of estimating the radiant flux and the mass eruption rate (e.g., [5,6]). Moreover, they enable the identification of thermal anomalies associated with hot degassing. This aspect is particularly relevant, considering that short-term variations of non-eruptive thermal fluxes are essential for volcano monitoring (e.g., [7]).

Among the systems developed to monitor active volcanoes from space, MODVOLC exploits MODIS (Moderate Resolution Imaging Spectroradiometer), MIR (Medium Infrared) and TIR (Thermal Infrared) data, at 1 km spatial resolution, to detect and quantify high-temperature features such as lava flows. MODVOLC provides information about thermal anomalies in terms of hotspot pixel number, total MIR radiance and radiant flux, in both nighttime and daytime conditions [8,9].

MIROVA [3] is another well-established MODIS-based system, performing over more than 300 active volcanoes [10]. This system shows a good efficiency in also detecting subtle hotspots (e.g., [3]). On the other hand, the lack of sensitivity of MODVOLC to low-level thermal anomalies, ascribable to the use of a fixed threshold test used globally, was demonstrated in several previous studies, also through comparison with MODLEN [11] and RST (Robust Satellite Techniques) (e.g., [12,13]). The latter is a multi-temporal scheme of satellite data analysis, which was used with success to study and monitor several active volcanoes located in different geographic areas, using high temporal resolution satellite data, from sensors such as AVHRR (Advanced Very High Resolution Radiometer), MODIS and SEVIRI (Spinning Enhanced Visible and Infrared Imager) (e.g., [14,15]), offering an extended dataset of multi-year satellite observations. The algorithm has also shown a high potential in detecting subtle thermal activities, which may precede volcanic eruptions (e.g., [2,4,16]).

In this work, we assess the RST potential in detecting and mapping volcanic thermal anomalies by satellite through ASTER (Advanced Spaceborne Thermal Emission and Reflection Radiometer) TIR data, at 90 m spatial resolution. ASTER was largely used in volcanology, despite the nominal revisit time of 16 days, thanks to its features in terms of spectral and spatial resolution (e.g., [17,18] and reference herein). Here, we show the results of the preliminary RST implementation on ASTER data, performed under the Google Earth Engine (GEE) environment. GEE is a cloud-based platform providing access to thousands of satellite images (e.g., European Space Agency-Sentinel, United States Geological Survey-Landsat), and enabling their handling and analysis at planetary scale (see [19] for more details). The high computational resources of GEE were previously exploited to develop an freely accessible tool devoted to map volcanic thermal anomalies at global scale (https://sites.google.com/view/nhi-tool), using MSI (Multispectral Instrument) and OLI (Operational Land Imager) SWIR (shortwave infrared) data at 20/30 m spatial resolution [20,21]. Here, we evaluate the contribution of an RST-based system analyzing ASTER TIR data, named RASTer (Robust ASTER Thermal anomaly system), in supporting the operational monitoring of active volcanoes from space, particularly for subtle hotspots identification. Mt. Etna (Italy), Shishaldin (AK, USA) and Shinmoedake (Japan) thermal activities are analyzed in this study by comparing RASTer detections to those from the largely accepted AVA (Aster Volcano Archive) database.

2. Materials and Methods

2.1. Advanced Spaceborne Thermal Emission and Reflection Radiometer (ASTER)

ASTER is one of five instruments (i.e., CERES, MISR, MODIS and MOPITT) aboard the Terra platform, launched in December 1999. Until April 2008, ASTER has collected images in 14 spectral bands (an additional backward-looking band is used for stereographic observations) in the VNIR (visible and near infrared), SWIR and TIR regions (Table 1). Since April 2008, ASTER has provided data only in the VNIR and TIR bands, owing to the failure of SWIR instrument. Each ASTER scene covers an area of 60×60 km; the spatial resolution ranges from 15 m (VNIR) to 90 m (TIR). Due to several constraints (see [22]), ASTER data acquisition is not everywhere guaranteed every 16 days, as its acquisitions follow a pre-defined and/or an "on-demand" scheduling.

Since November 2016, all ASTER Level 1 Precision Terrain Corrected Registered At-Sensor Radiance Product (AST_L1T) have been made freely available in Google Earth Engine. Data stored in AST_L1T imagery are calibrated at-sensor radiance, and they have been geometrically corrected and rotated to a north-up in UTM projection (see [23] for more details).

Table 1. Spatial and spectral resolutions of the ASTER bands.

Name	Resolution	Wavelength
B01	15 m	0.520–0.600 μm
B02	15 m	0.630–0.690 μm
B3N	15 m	0.780–0.860 μm
B04 *	30 m	1.600–1.700 μm
B05 *	30 m	2.145–2.185 μm
B06 *	30 m	2.185–2.225 μm
B07 *	30 m	2.235–2.285 μm
B08 *	30 m	2.295–2.365 μm
B09 *	30 m	2.360–2.430 μm
B10	90 m	8.125–8.475 μm
B11	90 m	8.475–8.825 μm
B12	90 m	8.925–9.275 μm
B13	90 m	10.250–10.950 μm
B14	90 m	10.950–11.650 μm

* Operational until April 2008.

2.2. Robust Satellite Techniques (RST)

The RST technique is an advanced change detection scheme (whose detailed description can be found in [24]), which considers each anomaly in space-time domain as a deviation from a "normal" state, which may be determined by analyzing multi-annual time series of homogeneous (e.g., same calendar month and overpass time) cloud-free satellite records. A statistically based index, named ALICE (Absolutely Llocal—double "l" is used to strengthen the concept that the index is local in space as well as in time [25]—Index of Change of the Environment [25]), taking into account the signal variability due to natural (e.g., different land covers, solar exposition) and observational (e.g., sun zenith angle, satellite view angles) conditions, enables the identification of anomalous events in the space-time domain, guaranteeing a high trade-off between reliability and sensitivity. The ALICE index, in its general formulation, is computed as follows:

$$\otimes_V (x,y,t) = \frac{V(x,y,t) - \mu_V(x,y)}{\sigma_V(x,y)} \tag{1}$$

where $V(x,y,t)$ is the signal measured at the time t and location (x,y), $\mu_V(x,y)$, and $\sigma_V(x,y)$ are corresponding expected value (generally the temporal mean) and standard deviation computed using a homogeneous dataset (see above).

The RST detection scheme was used in several previous studies to identify hotspots associated with volcanic activity (e.g., [2,4,12]), forest fires (e.g., [26]), gas flaring (e.g., [27]) and other events (e.g., [28]) analyzing MIR and/or TIR signal. Here, it is implemented, for the first time, on ASTER TIR data by assessing its performance in mapping thermal anomalies associated with volcanic activity.

2.3. Robust ASTER Thermal Anomaly System (RASTer)

In this paper, we implemented the RST algorithm under GEE by analyzing the signal measured in the ASTER channel 13, which is centered at 10.6 μm (see Table 1). The index in Equation (1) is then calculated pixel by pixel, according to the general RST prescriptions:

$$\otimes_{TIR} (x,y,t) = \frac{BT_{10.6}(x,y,t) - \mu_{BT_{10.6}}(x,y)}{\sigma_{BT_{10.6}}(x,y)} \tag{2}$$

In Equation (2), the role of the variable $V(x,y,t)$ is played by the brightness temperature $BT_{10.6}(x,y,t)$ measured at the Top of Atmosphere (TOA) at around 10.6 μm (band 13), $\mu_{BT_{10.6}}(x,y)$, and $\sigma_{BT_{10.6}}(x,y)$ are the temporal mean and standard deviation computed for each location (x,y) over 20 years (i.e., 2000–2020) of ASTER observations (e.g., a total of 635 ASTER images were analyzed to generate the spectral reference fields at Shinmoedake).

The $\otimes_{TIR}(x, y, t)$ index is a standardized variable having a Gaussian behavior. Hence, pixels having values of $\otimes_{TIR}(x, y, t) > 3$ have a probability around 99.85% of being anomalous (e.g., [13]). As discussed in previous studies, the ALICE index is intrinsically protected against site effects (e.g., natural warming of volcanic rocks [2]). On the other hand, in this work, due to a dataset much less populated than other sensors (in average 25–35 ASTER daytime/nighttime images per month, spanning over twenty years of observations, are available), we used a normalized index in combination with that in Equation (1) to also detect subtle hotspots with a high confidence level of detection:

$$ND_{B12-B13}(x, y, t) = \frac{BT_{9.1}(x, y, t) - BT_{10.6}(x, y, t)}{BT_{9.1}(x, y, t) + BT_{10.6}(x, y, t)} \tag{3}$$

In Equation (3), $BT_{9.1}(x, y, t)$ and $BT_{10.6}(x, y, t)$ are the brightness temperatures measured at ~9.1 μm (i.e., band 12) and ~10.6 μm (i.e., band 13). TOA (Top of Atmosphere) BT values were calculated from the radiance at sensor. High-temperature targets, emitting strongly at shorter rather than longer TIR wavelengths, should lead to positive values of the $ND_{B12-B13}$ index. Hence, RASTer identifies thermal anomaly if at least one of the following conditions is satisfied:

$$\begin{cases} \otimes_{TIR}(x, y, t) \geq 4 \ (high\ intensity\ hot\ spots) \\ OR \\ \otimes_{TIR}(x, y, t) \geq 3\ AND\ ND_{B12-B13}(x, y, t) > 0 \\ (mid - low\ intensity\ hot\ spots) \end{cases} \tag{4}$$

The first test enables the identification of high-temperature features such as lava flows, while the second one should favor the identification of less intense hotspots (e.g., those associated with a mid-low Strombolian activity), increasing the confidence level of detection by combining the two above defined indices (i.e., $\otimes_{TIR}(x, y, t)$ and $ND_{B12-B13}$).

2.4. The AVA Database

The ASTER Volcano Archive (AVA) is an archive of over more than 1500 active volcanoes dedicated to their global monitoring, providing access to thousands of daytime and nighttime ASTER imagery at full resolution. Moreover, the AVA database also includes information about spectral signatures of volcanic emissions (e.g., eruption columns and plumes), surficial deposits (e.g., lava flows) and eruption precursor phenomena (e.g., [29]). Since 2000, the AVA database has included products (e.g., thermal anomalies maps) generated from ASTER L1B granules. The AVA database, whose data and products are disseminated in a common format (e.g., HDF, TIFF, KML) through the website http://ava.jpl.nasa.gov was continuously updated until late 2017 ([18]). Information from this ASTER-based system was also used to assess thermal anomalies flagged by algorithms running on high temporal resolution satellite data (e.g., [30]). In this paper, we compare RASTer to AVA detections by quantifying differences in detecting and mapping thermal anomalies through ASTER TIR data over three different volcanic areas (see next section).

3. Results

3.1. Etna Volcano

Mt. Etna (37.748° N–14.999° E) is a stratovolcano located in Sicily (Italy) and represents one of the most active volcanoes in Europe. Gas/ash emissions, Strombolian activities, lava fountains and lava flows generally characterize its eruptive activity [31,32]. Mt. Etna eruptions generally occur from summit craters (i.e., Voragine, Bocca Nuova, North East Crater, Southeast Crater and New Southeast Crater; [33], e.g., during 2011–2015, [34–36], as well as at the time of writing, intense paroxysms occurred; [37]), although some voluminous flank eruptions were also recorded in recent years (e.g., July–August 2001; October 2002–January 2003; September 2004–March 2005; May 2008–July 2009; [38,39]).

Figure 1 displays the temporal trend of hot pixels flagged by RASTer (red triangles) and AVA (green triangles) at Mt. Etna over the period 2000–2017 by analyzing an area of 10 km of radius centered around the summit crater. The plot shows that both RASTer and AVA considered the eruptive events of July 2001 as the most intense (see hot pixels number). In general, all the flank eruptions (see boxes in Figure 1) occurring during the investigated period were well detected by both systems. Figure 2 displays, for instance, the thermal anomaly maps generated by RASTer during some Mt. Etna flank eruptions. In particular, the daytime ASTER scene of 29 July 2001 (Figure 2a) shows how RASTer mapped the areas inundated by lava (see red and yellow pixels), indicating that the lava flow, moving in the south direction, extended up to about 1120 m elevation. Figure 2b displays the RASTer detection of 18 March 2014, showing a lava overflow from summit craters.

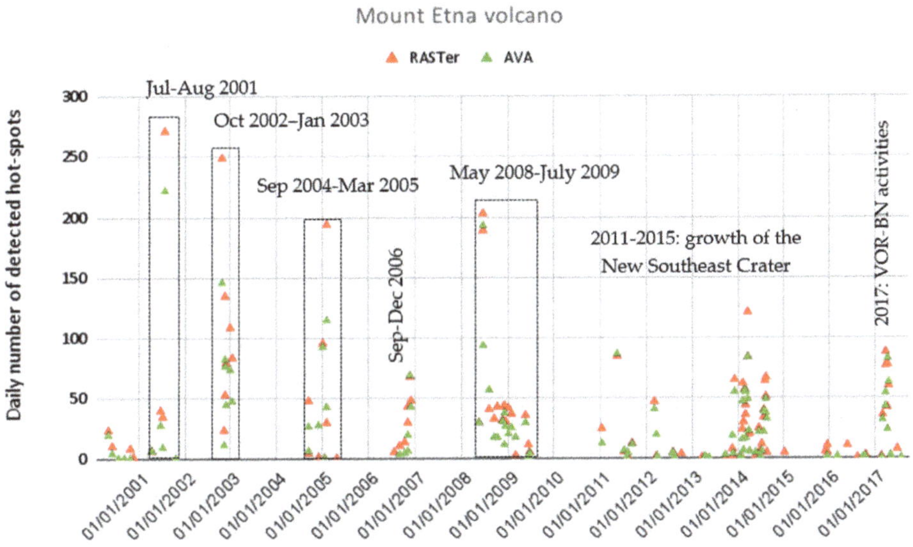

Figure 1. Time series (2000–2017) of the daily number of hot pixels flagged by RASTer and AVA at Mt. Etna. Dotted black rectangles indicate the periods of Mt. Etna flank eruptions; these eruptive events show the highest daily number of hotspots. The intense explosive and effusive activity at the Southeast Crater is also highlighted in the period September–December 2006. Similarly, the birth and growth of the New Southeast Crater in 2011–2015 is shown on the right side of the diagram. Finally, a significant number of hotspots characterized Strombolian activity in 2017, mainly at the Voragine (VOR) and Bocca Nuova (BN) craters.

In comparison with AVA, RASTer was capable of better mapping also less extended thermal anomalies (e.g., those associated with moderate Strombolian activities). Two examples are shown in Figure 3, displaying AVA and RASTer detections in reference to the Mt. Etna flank eruptions of 21 June 2008 and 27 October 2002. It is worth noting as RASTer was capable of mapping extensively and continuously (in the space domain) the main lava flows. Moreover, it correctly identified (on 27 October 2002, Figure 3b) some thermal anomalies that were not flagged at all by AVA (see crater area).

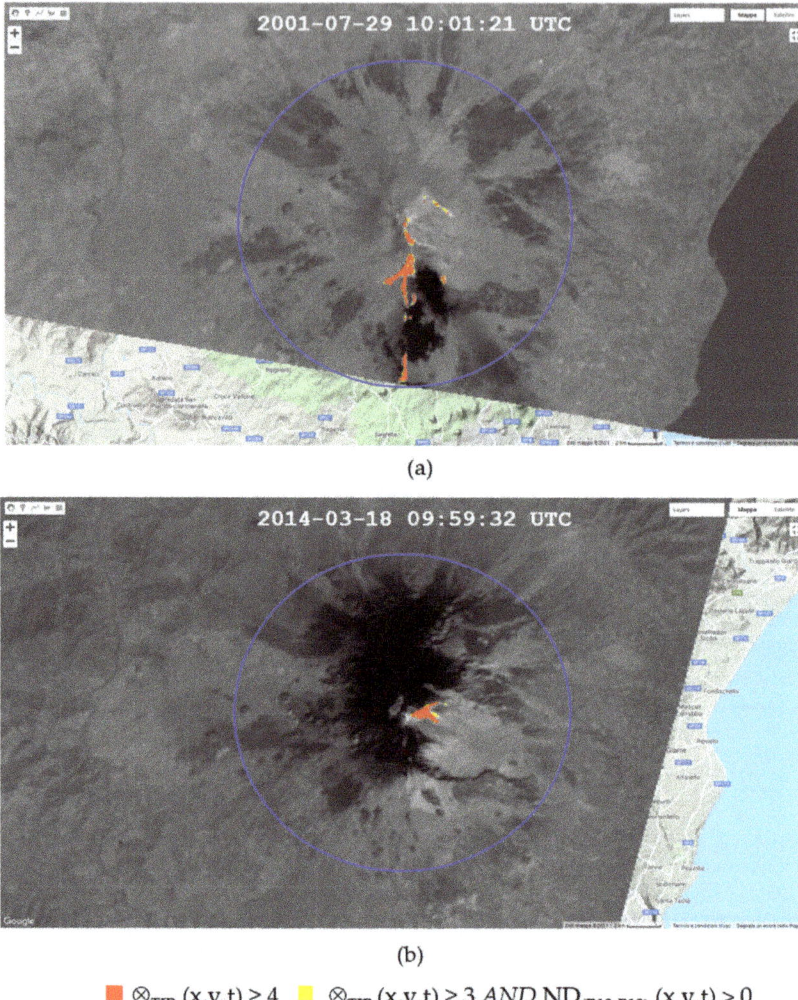

$\blacksquare$ $\otimes_{TIR}$ (x,y,t) ≥ 4 $\square$ $\otimes_{TIR}$ (x,y,t) ≥ 3 AND $ND_{(B12\text{-}B13)}$ (x,y,t) > 0

Figure 2. Thermal anomalies detected by RASTer on ASTER scenes during the Mt. Etna flank eruptions of (**a**) 29 July 2001; (**b**) 18 March 2014. More intense TIR anomalies referring to pixels with $\otimes_{TIR}(x,y,t) \geq 4$ are reported in red. The less intense ones, with $\otimes_{TIR}(x,y,t) \geq 3$ AND $ND_{B12-B13}(x,y,t) > 0$ (see text) are yellow colored. In the background, the ASTER band 3 TIR image.

The potential of RASTer in detecting subtle hotspots is emphasized in Figure 4, in reference to Strombolian activity of January–May 2020. The figure shows the identification of a thermal anomaly at the summit craters, where a Strombolian activity, gas-and-steam and ash emissions occurred since October 2019 (https://www.ct.ingv.it/, multidisciplinary bulletins), becoming spatially more extended during February–early March 2020. Afterward, the number of hot pixels decreased, marking the reduction of thermal volcanic activity at the Voragine-Bocca Nuova (VOR-BN) complex (the former Central Crater of Mt. Etna). The comparison with information from field observations demonstrates the capacity of the RASTer system in monitoring the whole dynamic of Mt. Etna thermal activity, up to its lowest levels of intensity.

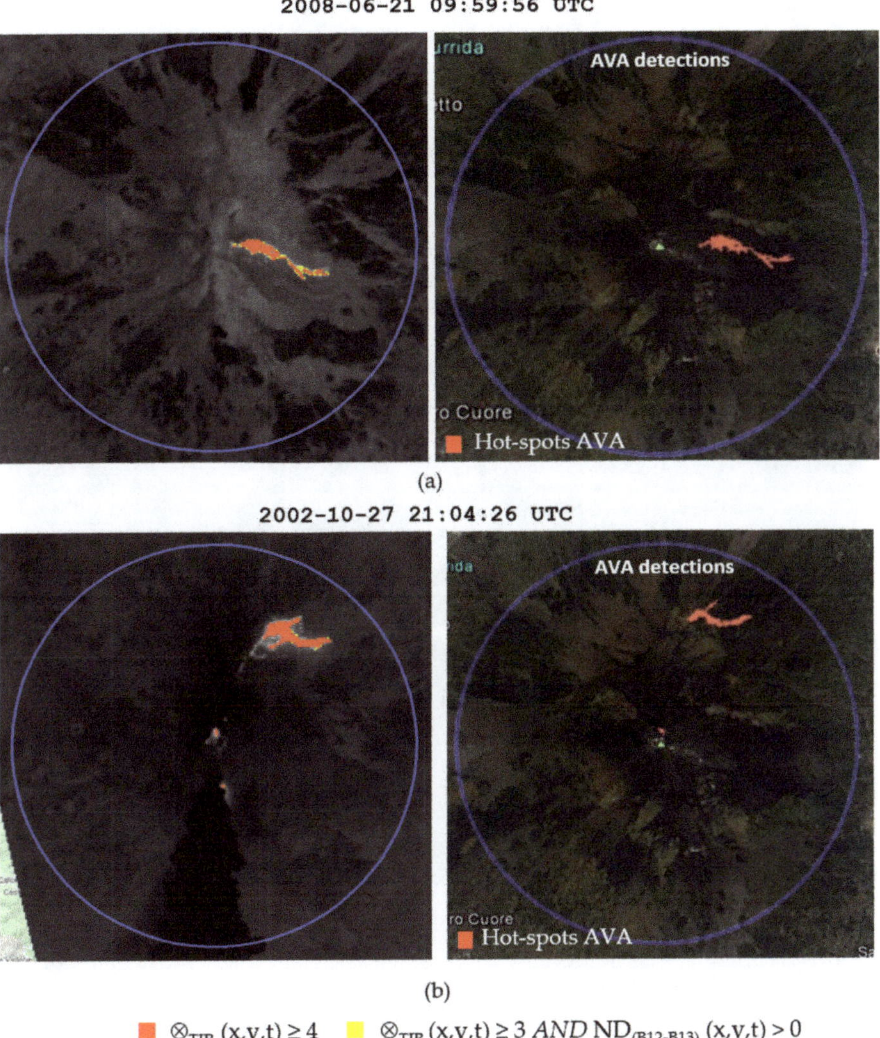

$\blacksquare$ $\otimes_{TIR}(x,y,t) \geq 4$ $\blacksquare$ $\otimes_{TIR}(x,y,t) \geq 3$ AND $ND_{(B12\text{-}B13)}(x,y,t) > 0$

Figure 3. Thermal anomalies detected on ASTER scenes during the Mt. Etna flank eruptions of (**a**) 21 June 2008 and (**b**) 27 October 2002 by RASTer (left side) and AVA (right side). Hotspots detected by RASTer are depicted as in Figure 2. The hotspots detected by AVA are depicted in red.

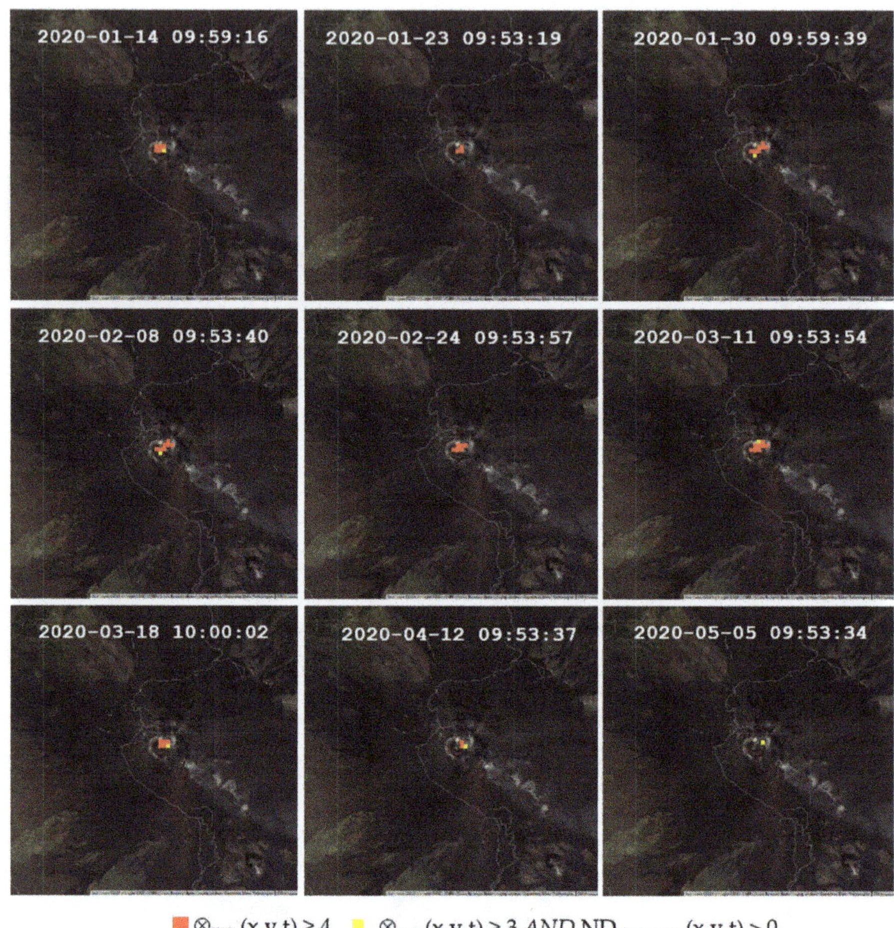

$\otimes_{\text{TIR}}(x,y,t) \geq 4$ $\otimes_{\text{TIR}}(x,y,t) \geq 3\ AND\ \text{ND}_{(\text{B12-B13})}(x,y,t) > 0$

Figure 4. Thermal anomalies detected by RASTer at Mt. Etna during the period January–May 2020, associated with the Strombolian activity recorded at the Voragine-Bocca Nuova complex.

3.2. Shishaldin Volcano

Shishaldin volcano is one of over 40 active volcanoes of Alaska (USA). This strato-volcano, which is located in the Unimak Island (54.7554° N–163.9711° W), has a nearly symmetrical cone (~16 km of diameter at the base) and reaches 2857 m above sea level (the highest altitude in the Aleutian Islands). About 40 historical eruptions have occurred at Shishaldin volcano (a summary of those events can be found in [40]). After 2000, a number of eruptive episodes occurred; the main eruption occurring during 2000–2017 was that recorded on 17 January 2014. It was characterized by the lava emission within the crater, low-level steam plumes, sporadic dustings of ash and ballistics on volcano flanks [41]. Starting from 30 January 2014, satellite observations revealed an increase of the surface temperature [42]. The presence of lava inside the crater was then reported on 25 March, 13 May, 1 October and in late January 2015 [40,42]. Afterwards, Shishaldin showed a low-level activity, continuing throughout 2015 and until the first months of 2016. A new lava emission inside the crater occurred on 23 October 2019. In the following weeks, lava flowed down the N and NE flanks; lava flows were revealed by independent satellite observations during November–December 2019 [43]. In January 2020, a new lava effusion occurred; lava affected the northeast (traveling down for 2 km) and north flanks, producing meltwater lahars [43].

Figure 5 displays the time series of RASTer and AVA detections at Shishaldin for the period January 2014–2016, when both systems were operational. We considered an area of 8000 m of radius around the summit crater to perform the comparison. The plot shows that RASTer detected a higher number of thermal anomalies than AVA and better emphasized a number of eruptive episodes of the investigated volcano (in terms of daily number of hotspots), indicating that the most intense Shishaldin activity occurred on 24 January 2015, in presence of lava inside the crater.

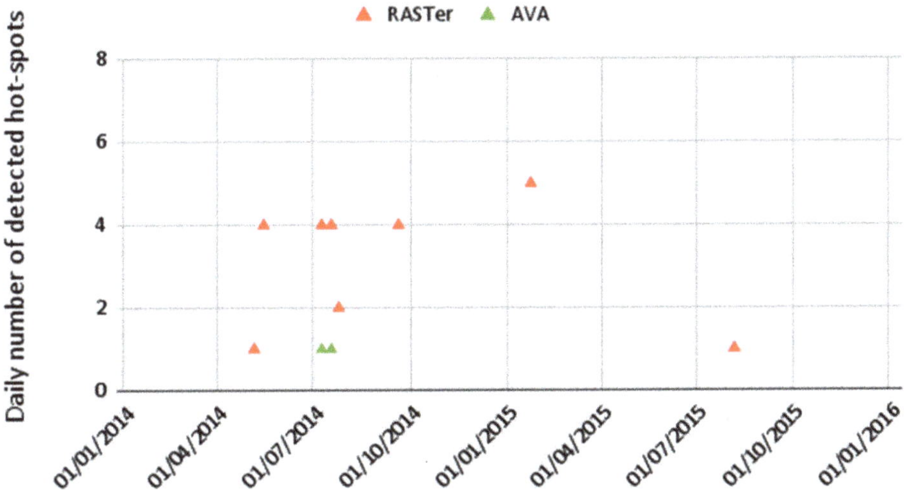

Figure 5. Time series of daily number of hotspots flagged by RASTer and AVA (January 2014–2016) at Shishaldin volcano.

Additionally, to assess the RASTer capacity in providing information about thermal anomalies at Shishaldin in recent years, Figure 6a shows an example of thermal anomalies detected on ASTER TIR data on 20 October 2019, in correspondence with a minor thermal activity that occurred at the summit crater. Figure 6b displays the lava flow on 8 January 2020, separating in two branches moving in the NW and NE direction, respectively. Those maps confirm the RASTer efficiency in mapping lava flows (providing unique information in periods when AVA products were not available), preserving a high sensitivity level even in areas located at the high latitudes, characterized by a cold background, where traditional fixed thresholds methods generally do not perform well (e.g., [44]).

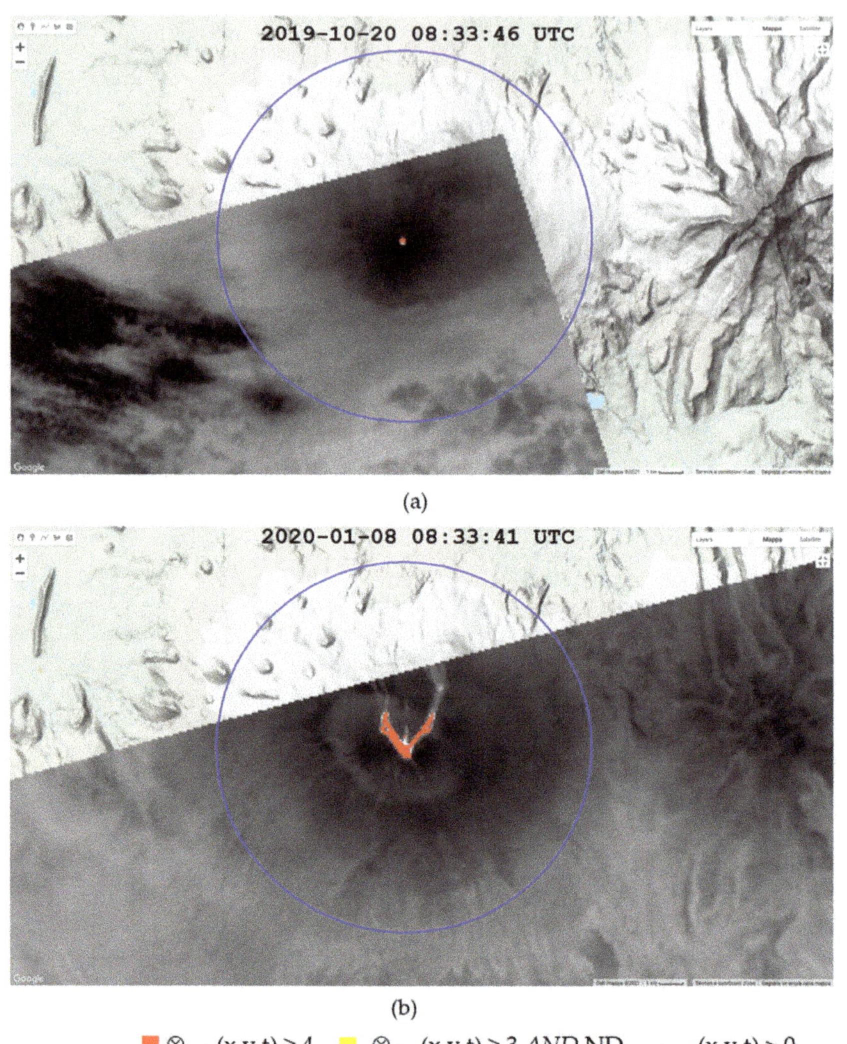

$\otimes_{\text{TIR}}(x,y,t) \geq 4$ $\otimes_{\text{TIR}}(x,y,t) \geq 3 \; AND \; \text{ND}_{(B12\text{-}B13)}(x,y,t) > 0$

Figure 6. Thermal anomalies detected by RASTer at Shishaldin volcano on ASTER scenes of 20 October 2019 (**a**) and 8 January 2020 (**b**).

3.3. Shinmoedake Volcano

Shinmoedake (31.931° N–130.864° E) is an andesitic stratovolcano (1421 m above sea level) located in Kyushu (Japan). The major Shinmoedake eruption occurred during 1716–1717. Other intense eruptive events were those of 1822, 1959, 1991 and 2008 (e.g., [45]). In recent years, a new eruption occurred on 19 January 2011 ([46]). On 28 January, a lava dome of about 200 m in diameter appeared on the crater floor. In the following three days, the dome grew up to 600 m in diameter, as indicated by independent satellite observations and overflights (e.g., [45,47]). The dome stopped growing in early February and only a number of explosive episodes occurred in the following months.

The comparison of RASTer and AVA detections at Shinmoedake revealed the occurrence of only two thermal anomalies over 17 years of ASTER observations. They were associated with the January–February 2011 eruption. Figure 7 shows these features, revealing that RASTer detected a higher number of hot pixels than AVA. In detail, Figure 7a shows the thermal anomalies associated with a lava dome inside the crater (on ASTER scenes of 31 January and 7 February 2011), which were underestimated by the AVA system (see Figure 7b).

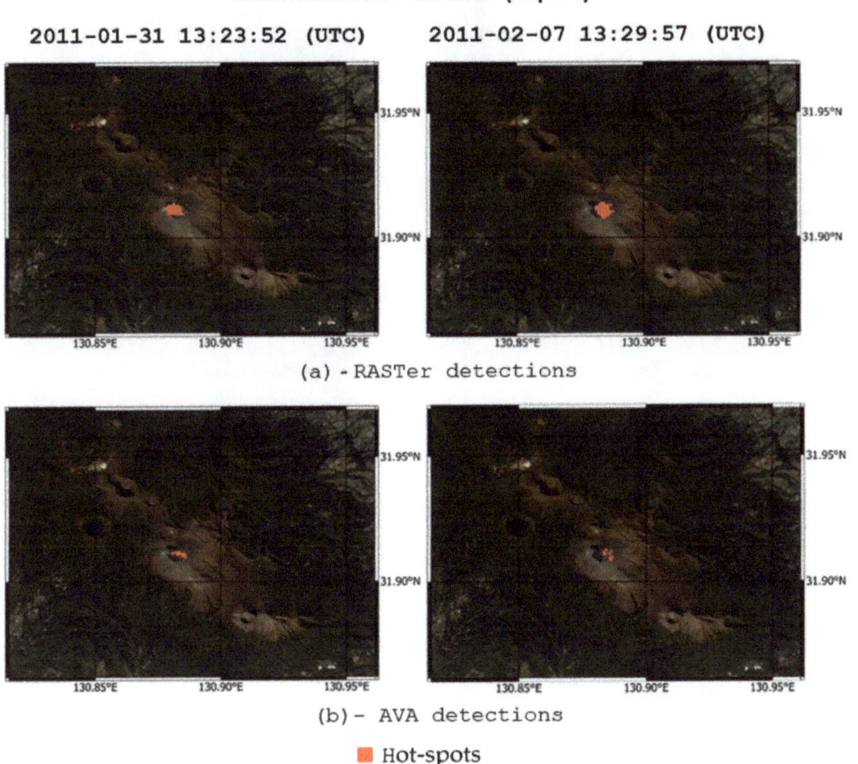

Figure 7. Thermal anomalies (red pixels) detected at Shinmoedake volcano on January–February 2011 (**a**) from RASTer; (**b**) from AVA.

After the eruptive episodes of 2011, a new eruption was recorded on 6 March 2018. A lava overflow to the NW rim of the summit crater was then observed three days later ([48]). Lava effusion stopped at the end of April 2018. Even in this case, it was not possible to compare RASTer with AVA detections. Figure 8 shows the thermal anomaly detected by RASTer from ASTER TIR data on 25 March 2018, confirming with a high accuracy level the capacity of the proposed system in mapping lava flows.

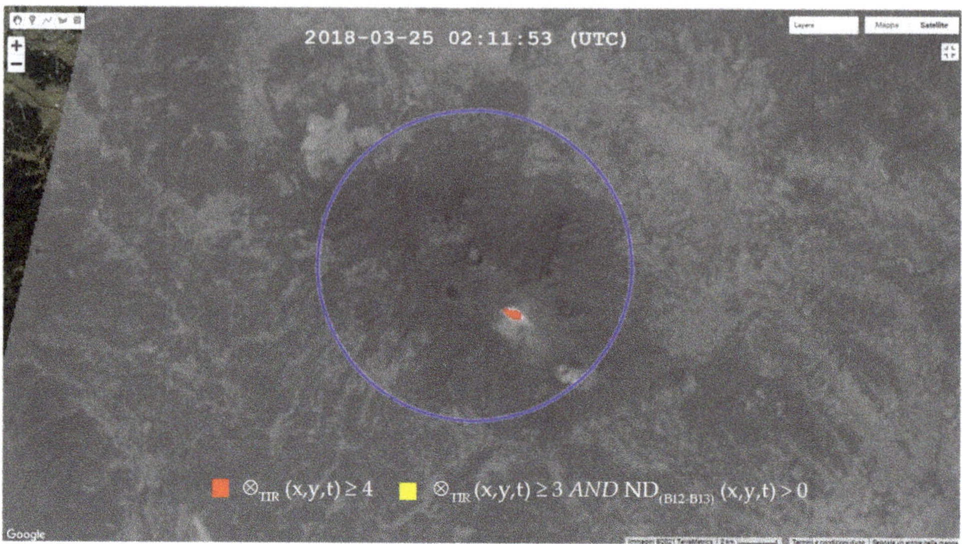

Figure 8. Thermal anomalies detected by RASTer at Shinmoedake volcano on 25 March 2018.

4. Discussion

We compared the RASTer and AVA detections in both nighttime and daytime conditions over three different active volcanic areas. Table 2 summarizes the results of this comparison, performed over the period May 2000–August 2017. The first column shows that at Mt. Etna and Shinmoedake, RASTer and AVA flagged the presence of thermal anomalies on the same number of ASTER scenes, although the number of hot pixels was significantly different (see second column). Common thermal anomaly detections characterized up to 86 ASTER scenes over Mt. Etna (see third column) where, however, higher was the number of hot pixels flagged by RASTer. On the other hand, the number of unique detections (fourth column) shows a possible complementarity of the two systems, which could increase the continuity of information. In general, the higher number of hotspots flagged by RASTer reveals a higher sensitivity to mid-low intensity thermal anomalies than AVA. This is also confirmed by results achieved at Shishaldin volcano, where only two thermal anomalies were flagged by AVA (see also results section).

Table 2. Summary of RASTer and AVA comparison performed by analyzing ASTER TIR data of 2000–2017 over Etna (Italy), Shinmoedake (Japan) and Shishaldin (AK, USA).

Volcano	Number of ASTER Images with Detected Thermal Anomalies (Until August 2017)		Number of Pixels Flagged as Thermal Anomalies		Common Detections			Unique Detections	
					Number of Scenes	Number of Hot Pixels		Number of Scenes	
	RASTer (R)	AVA (A)	R	A		R	A	R	A
Etna	100	100	4070	3036	86	3947	3017	14	14
Shinmoedake	2	2	60	19	2	60	19	0	0
Shishaldin	9	2	26	2	2	8	2	7	0

Thermal anomalies flagged by RASTer, including those undetected by AVA, were in good agreement with information from volcanological reports, as shown for instance in Figure 9 in reference to the Mt. Etna activity. The figure displays the thermal anomalies detected by RASTer on ASTER data of 2000–2020 and three different levels of thermal activity derived from field reports, independent observations and scientific papers [33–39,49,50]. It should be pointed out that RASTer detections were mostly associated with documented periods of Mt. Etna activity. An analysis is currently in progress to better assess hotspots flagged in periods of undocumented thermal volcanic activity (e.g., 27 July and 4 August 2012). Those features were probably associated with minor thermal activities (e.g., [12]) unreported by field reports, also confirming the relevance of satellite observations in providing information about subtle phases of unrest in well-monitored volcanic areas (e.g., [4]). Regarding the Shishaldin volcano—where, as for Shinmoedake, a long-quiescence period occurred, about 88.8% of detected thermal anomalies were ascribable to intra-crater thermal activities (e.g., May 2014–January 2015), and to lava effusions from volcano flanks (e.g., January 2020) reported by the Global Volcanism Program (GVP). Thermal anomaly detected by RASTer, ASTER scene of 12 August 2009, which was not confirmed by field reports, is consistent with MODIS observations of July–August 2009 [51].

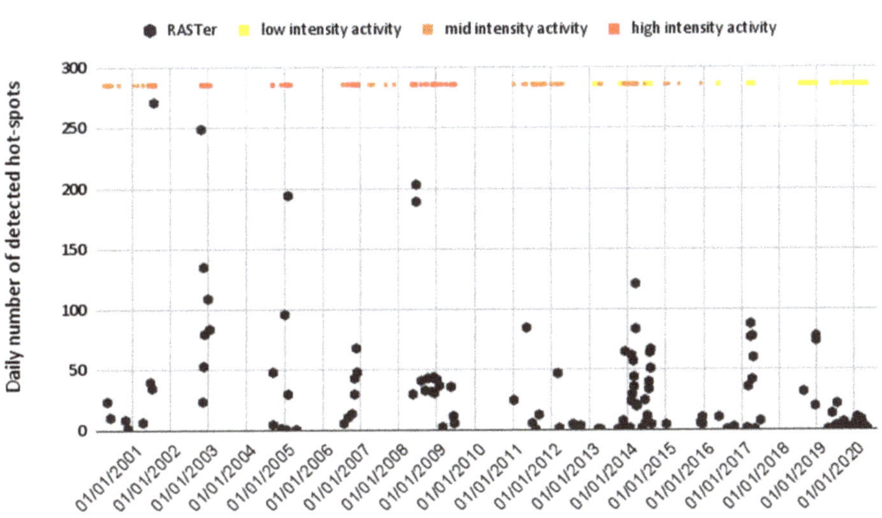

Figure 9. Time series of hot pixels flagged by RASTer at Mt. Etna in the period May 2000–2020 and different intensity levels of Mt. Etna thermal activity (colors from yellow to red).

Those results show that RASTer provided reliable information about volcanic thermal features of a different intensity level. This aspect is particularly relevant, considering that the RST technique for volcanological applications (e.g., RST_{VOLC} algorithm) was previously implemented on high temporal (low spatial) resolution satellite data, for which statistical analyses can be based on a highly populated dataset. Hence, this work demonstrates that RST is also capable of providing reliable results using mid-high spatial resolution satellite data and in the presence of a less populated dataset (as for ASTER). It should be stressed that, in this case, we did not implement the iterative pre-process used by RST to remove signal outliers (e.g., extremely hot/cold pixels) [24] within GEE. Nonetheless, RASTer was capable of performing an accurate mapping of thermal anomalies, as shown in Figure 10 in reference to the ASTER scene of 17 October 2016 at 09:59 UTC. In detail, Figure 10a shows the identification of some anomalous pixels at BN-VOR over the Mt. Etna crater area. This thermal anomaly detection integrated information retrieved from the Sentinel-2 MSI scene of the same day at 09:40 UTC, provided by the NHI (Normalized Hotspot Indices) algorithm through the GEE-based tool [20]. The latter, indeed, flagged some hot pixels over SEC (Southeast Crater-New Southeast Crater complex), (see blue pixels in Figure 10b), but did not provide information about the BN-VOR thermal activity detected by RASTer (red pixels). This thermal activity is probably associated with the presence of incandescent material under the fresh lava, which was independently observed on the ground a few days before the ASTER overpass (e.g., [52]). Differences in thermal anomaly identification are ascribable to the known limitations (e.g., missed detections in presence of degassing plumes [24]) and advantages (mainly associated with the use of satellite data at higher spatial resolution) of the NHI algorithm, which exploits NIR-SWIR Sentinel-2/MSI observations, when compared with the RST-based approach applied here to ASTER TIR data.

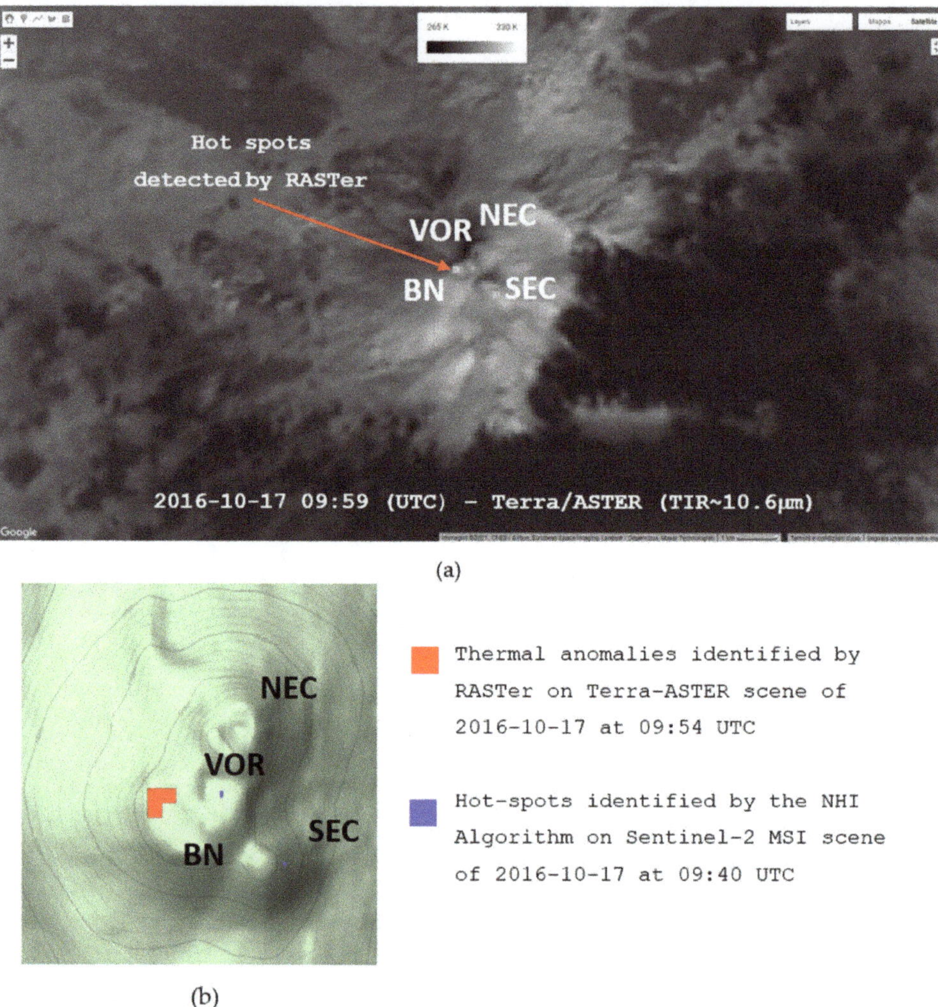

Figure 10. (**a**) ASTER TIR image 17 October 2016 at 09:54 UTC covering Mt. Etna area; (**b**) thermal anomalies detected by RASTer (red pixels) and NHI$_{SWIR}$ ([20]) on Sentinel-2/MSI data of 09:40 UTC (blue pixels) overlapped to the topographic map of Mt. Etna.

The possible complementarity between RASTer and NHI systems is further enhanced in the temporal domain, as shown in Figure 11 in reference to the Mt. Etna Strombolian activity of 14–17 January 2021 where the integration of ASTER and Sentinel-2 observations allowed for a more continuous monitoring of thermal volcanic activity. This integration can also be usefully extended to other sensors (e.g., NHI runs on Landsat-8/OLI data [20,21]), further increasing the temporal continuity of active volcanoes monitoring that, for the high-latitude regions, can today reach a mean temporal coverage up to 2 days (Figure 12).

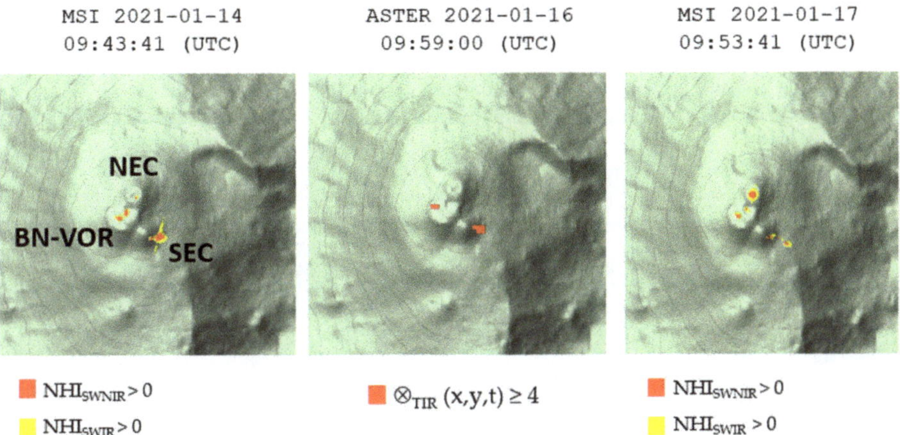

Figure 11. Mt. Etna activity detected by NHI through the indices NHI$_{SWNIR}$ and NHI$_{SWIR}$ (the latter is used to detect mid-low intensity hotspots [20]) on Sentinel-2/MSI data of 14 and 17 January 2021 (left and right panels) and by RASTer on ASTER scene of 16 January 2021 (middle panel).

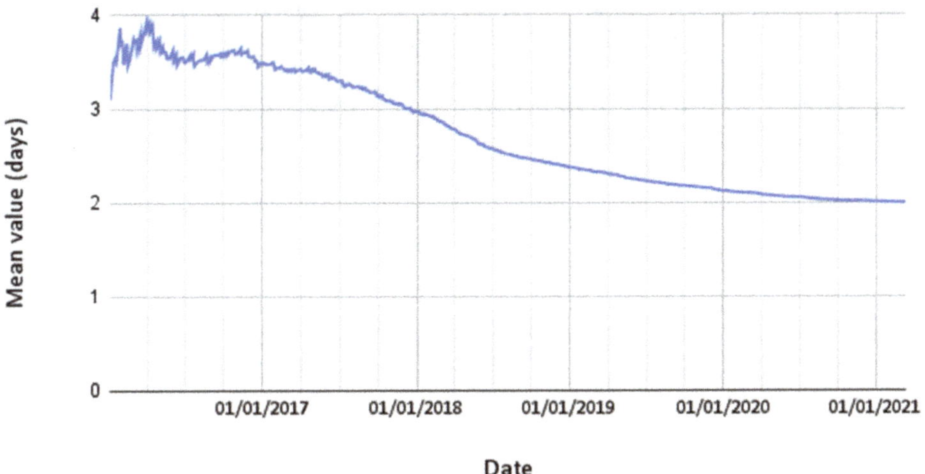

Figure 12. Evolution (2016–2021) of the temporal coverage achievable using a multi-satellite monitoring system integrating L8-OLI, S2-MSI and Terra-ASTER observations. The plot shows for the Shishaldin volcano location the average temporal gap among observations. Note as the launch of Sentinel-2B at the end of 2017 dramatically reduced the temporal gap from 4 on average (maximum 9) to 2 (maximum 3) days.

In addition, RASTer could be profitably used to assess information from systems monitoring volcanoes in near real time (using satellite data at lower spatial resolution), apart from areas (e.g., equatorial zones) where the scarcity and the bad quality of ASTER images do not enable the full RASTer implementation. Among those systems, the ones sharing the same RST-based approach, applied to data from polar (e.g., [2,14,30]) and geostationary (e.g., [53]) meteorological satellite sensors, will benefit from a RASTer-based validation and training. The RASTer contribution to detect and map lava flows and other thermal features related to volcanic activity is particularly relevant, considering that AVA products are no longer available (since 2017), the NHI usage on ASTER data is limited to 2000-2008 [54] and that other recent systems such as the ASTER Volcanic Thermal Output Database (AVTOD; [55]) operate over specific regions of interest and not globally. Indeed, ASTER also continues to play an important role in supporting volcanic activity monitoring from space at the time of writing. This is demonstrated by Figure 13, where lava flow (moving in E and SE direction toward Valle del Bove), associated with the powerful Mt. Etna paroxysm of 16 February 2021 and the thermal activity at summit craters were correctly identified and mapped by RASTer.

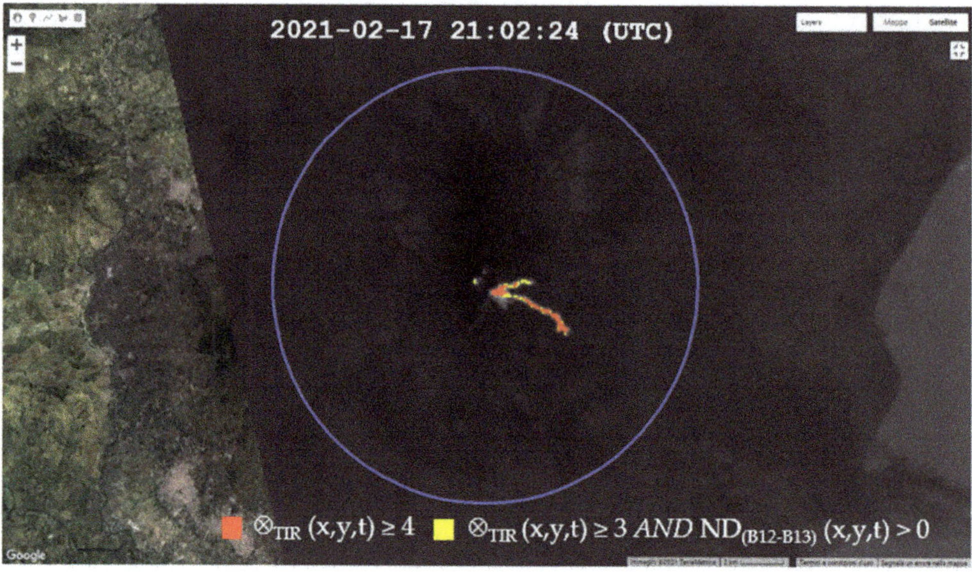

Figure 13. Thermal anomaly (red/yellow pixels) over Mt. Etna area detected by RASTer on 17 February 2021.

5. Conclusions

In this work, we presented and tested an RST-based algorithm to identify and map volcanic thermal features through ASTER TIR data, processed under the GEE platform. By analyzing results retrieved in three active volcanic areas (i.e., Etna, Shishadin and Shinmodake), through comparison with the AVA database, we found that RASTer was capable of successfully identifying thermal anomalies (as indicated by the relevant number of unique detections), also providing accurate information about these features in terms of shape and spatial extent. These outcomes demonstrate that RASTer may support the monitoring of thermal volcanic activity from space, despite some limitations. Main factors that could affect the RASTer detections are in fact bush fires, when occurring along the volcano flanks, and clouds that may obscure the underlying thermal anomalies. The computation of the normalized index defined in Equation (3) according to the general RST detection scheme, along with the implementation of an iterative procedure devoted to filter out signal outliers under the GEE environment, should further increase the RASTer performance and will be the aim of future investigations.

This study also encourages the RST implementation on data from previous Landsat missions and relative TIR sensors (Thematic Mapper and Enhanced Thematic Mapper plus). Those data should favor the investigation of thermal volcanic activities occurring in the past. Finally, this work opens new scenarios regarding the possible RST implementation on Landsat-8/TIRS (Thermal Infrared Sensor) data at 100 m spatial resolution, especially in view of the launch of the next Landsat-9 satellite, which will improve the temporal resolution of Landsat observations from 16 to 8 days.

Author Contributions: Conceptualization, N.G. and F.M.; software, N.G.; methodology, V.T., N.P. and N.G., validation and formal analysis, N.G., F.M. and M.N.; writing—original draft preparation, N.G. and F.M.; writing—review and editing, N.P., M.N. and V.T. All authors have read and agreed to the published version of the manuscript.

Funding: This research received no external funding.

Institutional Review Board Statement: Not applicable.

Informed Consent Statement: Not applicable.

Data Availability Statement: The data presented in this study are available under the Google Earth Engine platform.

Conflicts of Interest: The authors declare no conflict of interest.

References

1. Dehn, J.; Dean, K.; Engle, K. Thermal monitoring of North Pacific volcanoes from space. *Geology* **2000**, *28*, 755–758. [CrossRef]
2. Pergola, N.; Marchese, F.; Tramutoli, V. Automated detection of thermal features of active volcanoes by means of infrared AVHRR records. *Remote Sens. Environ.* **2004**, *93*, 311–327. [CrossRef]
3. Coppola, D.; Laiolo, M.; Cigolini, C. Fifteen years of thermal activity at Vanuatu's volcanoes (2000–2015) revealed by MIROVA. *J. Volcanol. Geotherm. Res.* **2016**, *322*, 6–19. [CrossRef]
4. Marchese, F.; Neri, M.; Falconieri, A.; Lacava, T.; Mazzeo, G.; Pergola, N.; Tramutoli, V. The Contribution of Multi-Sensor Infrared Satellite Observations to Monitor Mt. Etna (Italy) Activity during May to August 2016. *Remote Sens.* **2018**, *10*, 1948. [CrossRef]
5. Wright, R.; Pilger, E. Radiant flux from Earth's subaerially erupting volcanoes. *Int. J. Remote Sens.* **2008**, *29*, 6443–6466. [CrossRef]
6. Mastin, L.G. Testing the accuracy of a 1-D volcanic plume model in estimating mass eruption rate. *J. Geophys. Res. Atmos.* **2014**, *119*, 2474–2495. [CrossRef]
7. Connor, C.B.; Clement, B.M.; Song, X.; Lane, S.B.; West-Thomas, J. Continuous monitoring of high-temperature fumaroles on an active lava dome, Volcán Colima, Mexico: Evidence of mass flow variation in response to atmospheric forcing. *J. Geophys. Res. Space Phys.* **1993**, *98*, 19713–19722. [CrossRef]
8. Wright, R.; Flynn, L.; Garbeil, H.; Harris, A.; Pilger, E. Automated volcanic eruption detection using MODIS. *Remote Sens. Environ.* **2002**, *82*, 135–155. [CrossRef]
9. Wright, R.; Flynn, L.P.; Garbeil, H.; Harris, A.J.; Pilger, E. MODVOLC: Near-real-time thermal monitoring of global volcanism. *J. Volcanol. Geotherm. Res.* **2004**, *135*, 29–49. [CrossRef]
10. Coppola, D.; Laiolo, M.; Cigolini, C.; Massimetti, F.; Donne, D.D.; Ripepe, M.; Arias, H.; Barsotti, S.; Parra, C.B.; Centeno, R.G.; et al. Thermal Remote Sensing for Global Volcano Monitoring: Experiences from the MIROVA System. *Front. Earth Sci.* **2020**, *7*. [CrossRef]
11. Kervyn, M.; Ernst, G.G.J.; Harris, A.J.L.; Belton, F.; Mbede, E.; Jacobs, P. Thermal remote sensing of the low-intensity carbonatite volcanism of Oldoinyo Lengai, Tanzania. *Int. J. Remot. Sens.* **2008**, *29*, 6467–6499. [CrossRef]
12. Marchese, F.; Filizzola, C.; Genzano, N.; Mazzeo, G.; Pergola, N.; Tramutoli, V. Assessment and improvement of a robust satellite technique (RST) for thermal monitoring of volcanoes. *Remote Sens. Environ.* **2011**, *115*, 1556–1563. [CrossRef]
13. Lacava, T.; Marchese, F.; Pergola, N.; Tramutoli, V.; Coviello, I.; Faruolo, M.; Paciello, R.; Mazzeo, G. RSTVOLC implementa-tion on MODIS data for monitoring of thermal volcanic activity. *Ann. Geophys.* **2011**, *54*, 536–542. [CrossRef]
14. Pergola, N.; Coviello, I.; Filizzola, C.; Lacava, T.; Marchese, F.; Paciello, R.; Tramutoli, V. A review of RSTVOLC, an original algorithm for automatic detection and near-real-time monitoring of volcanic hotspots from space. *Geol. Soc. Lond. Spec. Publ.* **2015**, *426*, 55–72. [CrossRef]
15. Plank, S.; Marchese, F.; Filizzola, C.; Pergola, N.; Neri, M.; Nolde, M.; Martinis, S. The July/August 2019 Lava Flows at the Sciara del Fuoco, Stromboli–Analysis from Multi-Sensor Infrared Satellite Imagery. *Remote Sens.* **2019**, *11*, 2879. [CrossRef]
16. Marchese, F.; Lacava, T.; Pergola, N.; Hattori, K.; Miraglia, E.; Tramutoli, V. Inferring phases of thermal unrest at Mt. Asama (Japan) from infrared satellite observations. *J. Volcanol. Geotherm. Res.* **2012**, *237–238*, 10–18. [CrossRef]
17. Ramsey, M.S. Synergistic use of satellite thermal detection and science: A decadal perspective using ASTER. *Geol. Soc. Lond. Spec. Publ.* **2015**, *426*, 115–136. [CrossRef]
18. Ramsey, M.; Flynn, I. The Spatial and Spectral Resolution of ASTER Infrared Image Data: A Paradigm Shift in Volcanological Remote Sensing. *Remote Sens.* **2020**, *12*, 738. [CrossRef]
19. Gorelick, N.; Hancher, M.; Dixon, M.; Ilyushchenko, S.; Thau, D.; Moore, R. Google Earth Engine: Planetary-scale geospatial analysis for everyone. *Remote Sens. Environ.* **2017**, *202*, 18–27. [CrossRef]
20. Marchese, F.; Genzano, N.; Neri, M.; Falconieri, A.; Mazzeo, G.; Pergola, N. A Multi-Channel Algorithm for Mapping Volcanic Thermal Anomalies by Means of Sentinel-2 MSI and Landsat-8 OLI Data. *Remote Sens.* **2019**, *11*, 2876. [CrossRef]

21. Genzano, N.; Pergola, N.; Marchese, F. A Google Earth Engine Tool to Investigate, Map and Monitor Volcanic Thermal Anomalies at Global Scale by Means of Mid-High Spatial Resolution Satellite Data. *Remote Sens.* **2020**, *12*, 3232. [CrossRef]

22. Abrams, M.; Tsu, H.; Hulley, G.; Iwao, K.; Pieri, D.; Cudahy, T.; Kargel, J. The Advanced Spaceborne Thermal Emission and Reflection Radiometer (ASTER) after fifteen years: Review of global products. *Int. J. Appl. Earth Obs. Geoinf.* **2015**, *38*, 292–301. [CrossRef]

23. LP DAAC Team. Advanced Spaceborne Thermal Emission and Reflection Radiometer (ASTER) Level 1 Precision Terrain Corrected Registered At-Sensor Radiance Product (AST_L1T). Available online: https://lpdaac.usgs.gov/documents/300/ASTER_L1T_Product_Specification.pdf (accessed on 2 April 2021).

24. Tramutoli, V. Robust Satellite Techniques (RST) for Natural and Environmental Hazards Monitoring and Mitigation: Theory and Applications. In Proceedings of the IEEE Fourth International Workshop on the Analysis of Multitemporal Remote Sensing Images, Leuven, Belgium, 18–20 July 2007; pp. 1–6. [CrossRef]

25. Tramutoli, V. Robust AVHRR Techniques (RAT) for Environmental Monitoring: Theory and applications. In *Earth Surface Remote Sensing II*; Cecchi, G., Zilioli, E., Eds.; SPIE: Barcelona, Spain, 1998; Volume 349, pp. 101–113. [CrossRef]

26. Filizzola, C.; Corrado, R.; Marchese, F.; Mazzeo, G.; Paciello, R.; Pergola, N.; Tramutoli, V. RST-FIRES, an exportable algorithm for early-fire detection and monitoring: Description, implementation, and field validation in the case of the MSG-SEVIRI sensor. *Remote Sens. Environ.* **2017**, *192*, e2–e25. [CrossRef]

27. Faruolo, M.; Lacava, T.; Pergola, N.; Tramutoli, V. On the Potential of the RST-FLARE Algorithm for Gas Flaring Characterization from Space. *Sensors* **2018**, *18*, 2466. [CrossRef]

28. Filizzola, C.; Cerra, D.; Corrado, R.; De La Cruz, A.; Pergola, N.; Tramutoli, V. Robust Satellite Techniques (RST) for Pipeline Network Monitoring. In Proceedings of the IEEE Fourth International Workshop on the Analysis of Multitemporal Remote Sensing Images, Leuven, Belgium, 18–20 July 2007; pp. 1–5. [CrossRef]

29. Pieri, D.; Abrams, M. ASTER watches the world's volcanoes: A new paradigm for volcanological observations from orbit. *J. Volcanol. Geotherm. Res.* **2004**, *135*, 13–28. [CrossRef]

30. Lacava, T.; Kervyn, M.; Liuzzi, M.; Marchese, F.; Pergola, N.; Tramutoli, V. Assessing performance of the RSTVOLC multi-temporal algorithm in detecting subtle hot spots at Oldoinyo Lengai (Tanzania, Africa) for comparison with MODLEN. *Remote Sens.* **2018**, *10*, 1177. [CrossRef]

31. Branca, S.; Del Carlo, P. Types of eruptions of Etna volcano AD 1670–2003: Implications for short-term eruptive behaviour. *Bull. Volcanol.* **2005**, *67*, 732–742. [CrossRef]

32. Behncke, B.; Neri, M. Cycles and trends in the recent eruptive behaviour of Mount Etna (Italy). *Can. J. Earth Sci.* **2003**, *40*, 1405–1411. [CrossRef]

33. Acocella, V.; Neri, M.; Behncke, B.; Bonforte, A.; Del Negro, C.; Ganci, G. Why Does a Mature Volcano Need New Vents? The Case of the New Southeast Crater at Etna. *Front. Earth Sci.* **2016**, *4*, 67. [CrossRef]

34. Behncke, B.; Branca, S.; Corsaro, R.A.; De Beni, E.; Miraglia, L.; Proietti, C. The 2011–2012 summit activity of Mount Etna: Birth, growth and products of the new SE crater. *J. Volcanol. Geotherm. Res.* **2014**, *270*, 10–21. [CrossRef]

35. De Beni, E.; Behncke, B.; Branca, S.; Nicolosi, I.; Carluccio, R.; Caracciolo, F.D.; Chiappini, M. The continuing story of Etna's New Southeast Crater (2012–2014): Evolution and volume calculations based on field surveys and aerophotogrammetry. *J. Volcanol. Geotherm. Res.* **2015**, *303*, 175–186. [CrossRef]

36. Ganci, G.; Cappello, A.; Bilotta, G.; Hérault, A.; Zago, V.; Del Negro, C. Mapping Volcanic Deposits of the 2011–2015 Etna Eruptive Events Using Satellite Remote Sensing. *Front. Earth Sci.* **2018**, *6*. [CrossRef]

37. Zuccarello, F.; Schiavi, F.; Viccaro, M. Magma dehydration controls the energy of recent eruptions at Mt. Etna volcano. *Terra Nova* **2021**. [CrossRef]

38. Neri, M.; Casu, F.; Acocella, V.; Solaro, G.; Pepe, S.; Berardino, P.; Sansosti, E.; Caltabiano, T.; Lundgren, P.; Lanari, R. Deformation and eruptions at Mt. Etna (Italy): A lesson from 15 years of observations. *Geophys. Res. Lett.* **2009**, *36*. [CrossRef]

39. Neri, M.; Acocella, V.; Behncke, B.; Giammanco, S.; Mazzarini, F.; Rust, D. Structural analysis of the eruptive fissures at Mount Etna (Italy). *Ann. Geophys.* **2011**, *54*, 464–479. [CrossRef]

40. Nye, C.; Keith, T.; Eichelberger, J.; Miller, T.; McNutt, S.; Moran, S.; Schneider, D.J.; Dehn, J.; Schaefer, J.R. The 1999 eruption of Shishaldin Volcano, Alaska: Monitoring a distant eruption. *Bull. Volcanol.* **2002**, *64*, 507–519. [CrossRef]

41. Cameron, C.E.; Dixon, J.P.; Neal, C.A.; Waythomas, C.F.; Schaefer, J.R.; McGimsey, R.G. *2014 Volcanic Activity in Alaska—Summary of Events and Response of the Alaska Volcano Observatory*; U.S. Geological Survey Scientific Investigations Report 2017–5077; U.S. Geological Survey: Anchorage, AK, USA, 2017; 81p. [CrossRef]

42. Alaska Volcano Observatory. Shishaldin Reported Activity, Event Name: Shishaldin 2014/1. Available online: https://avo.alaska.edu/volcanoes/activity.php?volcname=Shishaldin&page=basic&eruptionid=771 (accessed on 2 April 2021).

43. Global Volcanism Program. *Bulletin of the Global Volcanism Network 45:2*; Report on Shishaldin (United, States); Krippner, J.B., Venzke, E., Eds.; Smithsonian Institution: Washington, DC, USA, 2020.

44. Lacava, T.; Marchese, F.; Arcomano, G.; Coviello, I.; Falconieri, A.; Faruolo, M.; Pergola, N.; Tramutoli, V. Thermal Monitoring of Eyjafjöll Volcano Eruptions by Means of Infrared MODIS Data. *IEEE J. Sel. Top. Appl. Earth Obs. Remote Sens.* **2014**, *7*, 3393–3401. [CrossRef]

45. Kato, K.; Yamasato, H. The 2011 eruptive activity of Shinmoedake volcano, Kirishimayama, Kyushu, Japan—Overview of activity and Volcanic Alert Level of the Japan Meteorological Agency. *Earth Planets Space* **2013**, *65*, 489–504. [CrossRef]

46. Global Volcanism Program. *Bulletin of the Global Volcanism Network 35:12*; Report on Kirishimayama (Japan); Wunderman, R., Ed.; Smithsonian Institution: Washington, DC, USA, 2010.

47. Ozawa, T.; Kozono, T. Temporal variation of the Shinmoe-dake crater in the 2011 eruption revealed by spaceborne SAR observations. *Earth Planets Space* **2013**, *65*, 527–537. [CrossRef]

48. Global Volcanism Program. *Bulletin of the Global Volcanism Network 43:6*; Report on Kirishimayama, (Japan); Crafford, A.E., Venzke, E., Eds.; Smithsonian Institution: Washington, DC, USA, 2018. [CrossRef]

49. Neri, M.; De Maio, M.; Crepaldi, S.; Suozzi, E.; Lavy, M.; Marchionatti, F.; Calvari, S.; Buongiorno, M.F. Topographic Maps of Mount Etna's Summit Craters, updated to December 2015. *J. Maps* **2017**, *13*, 674–683. [CrossRef]

50. De Beni, E.; Cantarero, M.; Neri, M.; Messina, A. Lava flows of Mt Etna, Italy: The 2019 eruption within the context of the last two decades (1999–2019). *J. Maps* **2020**, 1–12. [CrossRef]

51. Global Volcanism Program. *Bulletin of the Global Volcanism Network 40:1*; Report on Shishaldin (United States); Wunderman, R., Ed.; Smithsonian Institution: Washington, DC, USA, 2015. [CrossRef]

52. Global Volcanism Program. *Bulletin of the Global Volcanism Network 42:9*; Report on Etna, (Italy); Crafford, A.E., Venzke, E., Eds.; Smithsonian Institution: Washington, DC, USA, 2017. [CrossRef]

53. Marchese, F.; Falconieri, A.; Pergola, N.; Tramutoli, V. A retrospective analysis of the Shinmoedake (Japan) eruption of 26–27 January 2011 by means of Japanese geostationary satellite data. *J. Volcanol. Geotherm. Res.* **2014**, *269*, 1–13. [CrossRef]

54. Mazzeo, G.; Ramsey, M.S.; Marchese, F.; Genzano, N.; Pergola, N. Implementation of the NHI (Normalized Hot Spot Indices). Algorithm on Infrared ASTER Data: Results and Future Perspectives. *Sensors* **2021**, *21*, 1538. [CrossRef] [PubMed]

55. Reath, K.; Pritchard, M.E.; Moruzzi, S.; Alcott, A.; Coppola, D.; Pieri, D. The AVTOD (ASTER Volcanic Thermal Output Da-tabase) Latin America archive. *J. Volcanol. Geotherm. Res.* **2019**, *376*, 62–74. [CrossRef]

Article

Copernicus Sentinel-1 MT-InSAR, GNSS and Seismic Monitoring of Deformation Patterns and Trends at the Methana Volcano, Greece

Theodoros Gatsios [1,2], Francesca Cigna [3,*], Deodato Tapete [3], Vassilis Sakkas [1], Kyriaki Pavlou [1] and Issaak Parcharidis [2]

[1] Department of Geophysics and Geothermy, National and Kapodistrian University of Athens (NKUA), Panepistimiopolis-Zographou, 15784 Athens, Greece; theogat@geol.uoa.gr (T.G.); vsakkas@geol.uoa.gr (V.S.); kpavlou@geol.uoa.gr (K.P.)
[2] Department of Geography, Harokopio University of Athens (HUA), 70 El. Venizelou Str., 17671 Athens, Greece; parchar@hua.gr
[3] Italian Space Agency (ASI), Via del Politecnico snc, 00133 Rome, Italy; deodato.tapete@asi.it
* Correspondence: francesca.cigna@asi.it

Received: 23 August 2020; Accepted: 14 September 2020; Published: 16 September 2020

Abstract: The Methana volcano in Greece belongs to the western part of the Hellenic Volcanic Arc, where the African and Eurasian tectonic plates converge at a rate of approximately 3 cm/year. While volcanic hazard in Methana is considered low, the neotectonic basin constituting the Saronic Gulf area is seismically active and there is evidence of local geothermal activity. Monitoring is therefore crucial to characterize any activity at the volcano that could impact the local population. This study aims to detect surface deformation in the whole Methana peninsula based on a long stack of 99 Sentinel-1 C-band Synthetic Aperture Radar (SAR) images in interferometric wide swath mode acquired in March 2015–August 2019. A Multi-Temporal Interferometric SAR (MT-InSAR) processing approach is exploited using the Interferometric Point Target Analysis (IPTA) method, involving the extraction of a network of targets including both Persistent Scatterers (PS) and Distributed Scatterers (DS) to augment the monitoring capability across the varied land cover of the peninsula. Satellite geodetic data from 2006–2019 Global Positioning System (GPS) benchmark surveying are used to calibrate and validate the MT-InSAR results. Deformation monitoring records from permanent Global Navigation Satellite System (GNSS) stations, two of which were installed within the peninsula in 2004 (METH) and 2019 (MTNA), are also exploited for interpretation of the regional deformation scenario. Geological, topographic, and 2006–2019 seismological data enable better understanding of the ground deformation observed. Line-of-sight displacement velocities of the over 4700 PS and 6200 DS within the peninsula are from −18.1 to +7.5 mm/year. The MT-InSAR data suggest a complex displacement pattern across the volcano edifice, including local-scale land surface processes. In Methana town, ground stability is found on volcanoclasts and limestone for the majority of the urban area footprint while some deformation is observed in the suburban zones. At the Mavri Petra andesitic dome, time series of the exceptionally dense PS/DS network across blocks of agglomerate and cinder reveal seasonal fluctuation (5 mm amplitude) overlapping the long-term stable trend. Given the steepness of the slopes along the eastern flank of the volcano, displacement patterns may indicate mass movements. The GNSS, seismological and MT-InSAR analyses lead to a first account of deformation processes and their temporal evolution over the last years for Methana, thus providing initial information to feed into the volcano baseline hazard assessment and monitoring system.

Keywords: SAR; InSAR; ground deformation; Sentinel-1; volcano monitoring; GNSS; seismicity; ground deformation; slope instability; MT-InSAR

---

1. Introduction

It is well known that some of the most scenic landscapes and islands of Greece are due to the intense volcanic activity that occurred centuries to millennia ago [1,2]. While Santorini is the most famous and a worldwide renowned touristic site, there are other active volcanic systems (i.e., Milos and Nisyros) within the south Aegean volcanic arc, which were formed from the subduction of the African tectonic plate beneath the Eurasian plate. Of this arc, the Methana peninsula is the westernmost dormant but geodynamically and hydrothermally active volcanic system [3]. The peninsula is connected with the north-eastern coast of Peloponnesus by a narrow gooseneck-shaped isthmus and stretches for approximately 44 km^2 northward into the Saronic Gulf.

Of the around 32 andesitic and dacitic lava domes scattered across the peninsula, the Methana volcano is the largest and is the subject of the present study. The volcanic hazard in Methana is considered "low" [4]; given that the last historic eruption was registered in approximately 230 BC, volcanic products derived from highly explosive eruptions were not found [3] and no alarming signs were observed in recent times. However, several aspects provide sufficient motivation for a dedicated investigation into the Methana volcano. From a hazard point of view, the neotectonic basin constituting the Saronic Gulf area is considered seismically active [5], and the active fault systems therein were considered preferential paths for present-day geothermal fluid leakage and, as such, potential sites for magma uprising [3]. Thermal springs are indeed clear manifestations of volcanic geothermal energy in Methana, and gas exhalations suggest a mixture between a dominant hydrothermal component and mantle-derived fluids [3]. Additionally, from a risk point of view, the volcano is not far from Athens (less than 50 km south-west) and any activity at the volcano would impact the local population (about 2500 inhabitants over the year, increasing during summer due to tourism). Further proof that monitoring the Methana volcano has recently been given higher priority on the agenda of Greek institutions is provided by the installations of six seismological stations by the Greek Institute of Geodynamics of the National Observatory of Athens (NOA) [6] alongside a new NOANET Global Navigation Satellite System (GNSS) station [7]. This recent development significantly updates what was stated by [4], i.e., of the whole Greek volcanic arc of Methana–Milos–Santorini (Thera)–Nisyros, Methana is among the volcanic fields not efficiently monitored.

In this context, the present study aims to investigate and characterize the ground motions observed at the surface of Methana volcano based on the results achieved by means of multi-interferogram processing of Synthetic Aperture Radar (SAR) images collected by the Copernicus Programme Sentinel-1 satellite constellation in the period from March 2015 to August 2019. Interferometric SAR (InSAR) approaches have been largely exploited to study and monitor surface deformation at different locations across the volcanic arc (e.g., [8–17]) and have proved their effectiveness to provide a spatially distributed estimation of volcanic activity due to magma chamber processes as well as shallow deformation associated with hydrothermal activity (e.g., low temperature venting). Despite this abundant literature, to the best of our knowledge, no InSAR study has previously focused on Methana.

Given the varied land cover and presence of vegetation at Methana, we intentionally selected a Multi-Temporal InSAR (MT-InSAR) processing approach providing both Persistent Scatterers (PS) and Distributed Scatterers (DS) in order to augment the monitoring capability and the number of measurement points. In this way, we aimed to characterize not only the geodynamics of the whole volcano but also the surface deformation along the steep slopes and stream gullies of the rugged terrain of the volcanic landforms, of which the susceptibility to landslides and rockfalls has been already highlighted in the literature [18]. Satellite MT-InSAR is complemented and interpreted in combination with an analysis of regional seismicity characterizing the 2006–2019 period and with geodetic data from continuous GNSS monitoring and Global Positioning System (GPS) benchmark surveying. The latter is also used to calibrate and validate the MT-InSAR results, thus tying the satellite-based estimates to the local geodetic reference used in the region and providing evidence to corroborate MT-InSAR observations.

2. The Study Area

Methana peninsula is located in the western Saronic Gulf south of Athens city (Figure 1) and is one of the three volcanic centres (i.e., Aegina, Poros and Methana) representing the north-western part of the active volcanic arc of the South Aegean. This Hellenic active volcanic arc also includes the islands Milos, Santorini and Nisyros in its extension [19]. Methana and the other volcanic centres in the Saronic Gulf are mostly monogenetic, and no composite volcanic structures are present [20].

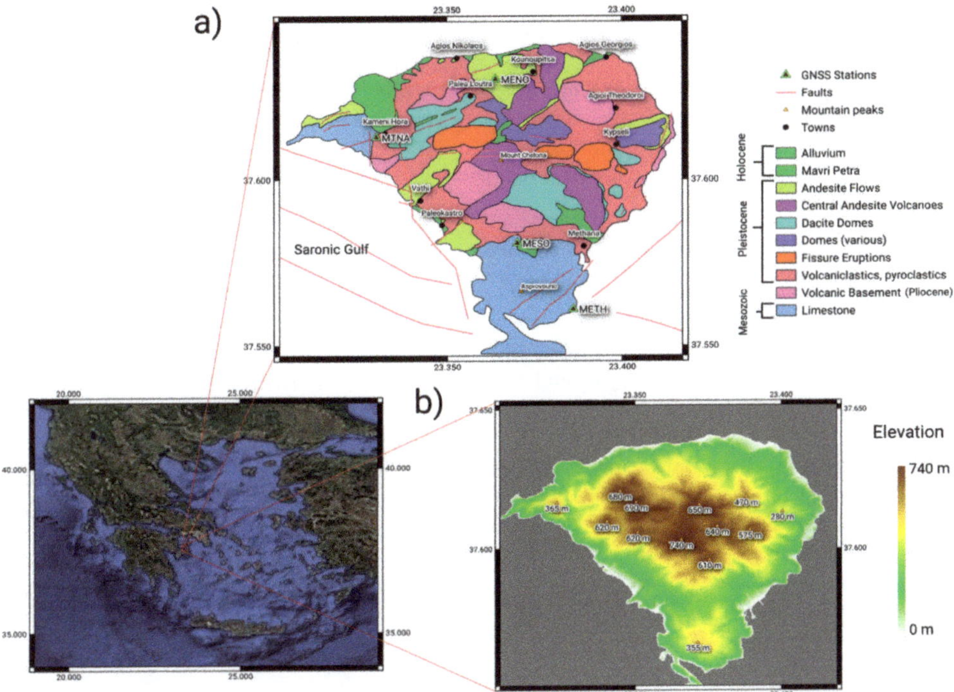

Figure 1. (**a**) Simplified geological map of the Methana peninsula, Greece (modified after [21–23]), with indication of Global Navigation Satellite System (GNSS) stations and local benchmarks and (**b**) topography from a 5.5-m resolution digital surface model by the Hellenic Mapping and Cadastral Organisation.

According to [24], the Saronic Gulf is the longest active area of the South Aegean with geothermal springs which are influenced by the active tectonic structure of the area as well as the existence of the magmatic chamber. Volcanic activity in Methana lasted until 1700, when a submarine eruption occurred north of Kameni Chora. Present day, active geothermal springs are reported mainly in the eastern and northern parts of the peninsula [3]. Within Methana town, in the south-east, the thermal springs are exploited at the open-air baths of the Thermal Spa and its neoclassical dwelling built in 1930.

It is also crucial to mention the presence of the Pausanias submarine volcano, located 2 km off the north-western shore of the peninsula [25]. According to [26], this submarine volcano was discovered using seismic reflection data and volcanic activity occurred during the 3rd century BC.

The Saronic Gulf constitutes a neotectonic basin which was created 4–5 million years ago [27,28]. NE-SW normal faults dominate the broader region, while N-S, E-W and NNE-SSW striking normal faults are also recognizable in the area [25] and are characterized by low seismic activity, chiefly with shallow events [29].

The geology of Methana consists of various volcanic formations with Late Pliocene–Pleistocene small andesitic to dacitic flows and domes [30,31], which overlay the older volcanic rocks, as well as the Mesozoic sedimentary basement [21,32].

A simplified geological map (Figure 1a) shows that the grey limestones are present in the northwest (Upper Triassic–Lower Jurassic), while limestones with dolomites are present in the south (Upper Jurassic–Cretaceous). These limestones constitute the basement of Methana and are located under the volcanoclastic formations (rhyolite, dacite and andesite) [33]. Additionally, the Methana peninsula has many tectonic faults, with a major fault crossing the town of Methana from west to east.

The tectonic regime of the region combined with the volcanic activity determined a rough topography of the peninsula, which shows both flat basin areas with Quaternary fill and zones dominated by high slope gradients and narrow valleys, often modelled by erosion processes (such as in the south, between Vathi and Methana) [18]. The maximum elevation is 740 m above sea level (a.s.l.) and occurs at Helona Mountain (Figure 1b).

3. Materials and Methods

3.1. SAR Data and Multi-Temporal InSAR Processing

A long stack of 99 SAR scenes acquired by the Copernicus Sentinel-1A and Sentinel-1B satellites in ascending mode (relative orbit 102) was used in this study. SAR images are remote sensing satellite data increasingly used for observing, mapping and monitoring Earth's surface processes [34] and are characterized by cloud-penetrating, day and night operational capabilities.

Among the Sentinel-1 SAR constellation imaging modes, we used the Interferometric Wide (IW) swath, which is based on the novel Terrain Observation with Progressive Scans (TOPS) strategy [35]. IW images are composed of three sub-swaths (i.e., IW1, IW2 and IW3), each consisting of a series of bursts, and provide a 250-km large coverage, with pixel resolutions of 5 m and 20 m (single look) in range and azimuth, respectively [36]. The Sentinel-1 data used in this study were collected from the Sentinel Hub platform and cover the time period from March 2015 to August 2019. The ascending mode acquisition geometry was selected to undertake the analysis in order to optimize visibility of the east-facing slopes, where the town of Methana (and thus the major urban settlement of the peninsula) is located (see Figure 1).

The stack of 99 Sentinel-1 SAR scenes was processed using an advanced MT-InSAR approach, an extension of the basic technique of Differential InSAR (DInSAR; e.g., [37,38]), that is based on the phase comparison of multiple SAR images gathered at different times over the study region and allows for measurements of land deformation along the line-of-sight (LOS) direction of the SAR sensor with up to millimetre precision (e.g., [39]). Advanced DInSAR methods are increasingly being used in various applications in the field of volcanology, seismology, crustal dynamics, landslides, land subsidence and geothermal energy exploitation (e.g., [40–45]).

MT-InSAR methods exploit either Persistent (PS) or Distributed (DS) Scatterers or coherent targets. Typically, PS are artificial objects that reflect the radar signal well, such as metal structures, buildings and rock outcrops, and are used in Persistent Scatterer Interferometry (PSI), e.g., the Permanent Scatterers (PS-InSAR) technique [46,47]. In urban areas, there is a prevalence of PS, and PSI methods allow for analysis of even individual structures on the ground. Methods exploiting coherent targets include algorithms such as the Small BAseline Subset (SBAS) method [48]. DS reflects lower radar energy compared to PS targets and usually spans several pixels in high-resolution SAR images [49], which exhibit similar scattering properties and can be used together for deformation estimation. The identification and monitoring of such targets is helpful especially in suburban and more rural areas, where the density of PS can be low. In such circumstances, the combination of PS and DS is important for effective displacement monitoring using SAR Single Look Complex (SLC) data. What is certain is that the processing aiming to detect both PS and DS requires large computing resource and processing time [50].

Concerning the precision of MT-InSAR measurements, error bars are extremely complex to estimate theoretically, since they depend on several factors, including the number of SAR SLC images used, the spatial density and quality (signal-to-noise ratio levels) of the measurement targets, their distance from the reference PS, the climatic conditions at the time of SAR acquisition and the satellite repeat cycle. Recent studies provide an empirical relationship between the number of SAR interferograms used and the resulting precision of the ground deformation estimates (e.g., by using 50 ERS-1/2 interferograms, a precision of 0.15 cm/year can be achieved with the SBAS method [51]).

In this study, the GAMMA SAR and Interferometry software (GAMMA Remote Sensing AG, Gümligen, Switzerland) and the Interferometric Point Target Analysis (IPTA) method [52] were used for the interferometric processing. The MT-InSAR processing workflow can be divided into two steps: the pre-processing and the main processing using GAMMA/IPTA.

During pre-processing, parameter files for the 99 SLC images were created and used to perform orbital refinement. The SLC images were co-registered based on the image used as a reference (20/09/2017), and a single-reference stack of 98 differential interferograms was generated. The Shuttle Radar Topography Mission (SRTM) 1-arcsec (i.e., 30 m/pixel) Digital Elevation Model (DEM) [53] was used during pre-processing to subtract the topography contribution from the interferograms.

Using the 99 co-registered SLC images, a multi-reference stack was also created, containing both single-pixel and multi-look differential interferometric phases. Multi-look differential interferometric phases (DS) were first generated, and afterwards, single-pixel differential interferometric phases (PS) were estimated using low spectral diversity and temporal variability of SLC intensity. Then, single-look and multi-look phases were combined into one list of both PS and DS. The multi-reference stack consists of 291 differential interferograms, pairing each scene with the 3 subsequent scenes (triple redundancy). The perpendicular baselines of those 291 pairs were quite short, with the longest baseline being 202 m. A high-accuracy Digital Surface Model (DSM) (5.5 m/pixel) provided by the Hellenic Mapping and Cadastral Organisation (Figure 1b) was also used for the IPTA in order to improve the quality of the output results. As reference point, the closest point to the GPS benchmark "Methana South" (i.e., MESO station in Figure 1a; 37°34′51.02562″ N, 23°22′12.14638″ E) was selected. Then, height corrections and atmospheric phases were computed using the combined point list.

The above model was refined by updating height corrections and atmospheric phases via iteration of the last step two more times. Finally, the phases were converted into deformation time series, which contained both PS and DS. Then, the deformation time series database was split into two separate ones, one for the PS and the other for the DS targets.

3.2. GNSS Continuous Monitoring Data and GPS Benchmark Surveying

A small network of two benchmark stations was established in 2006 in Methana Peninsula (Figure 1) [54]. These two stations were part of a larger GPS network established in the north-western part of the Hellenic Volcanic Arc (including the island of Aegina and the volcanic area of Susaki) aiming to study the ground deformation of this area. The stations in Methana were located in the northern (i.e., station MENO) and southern (MESO) parts of the peninsula and were reoccupied several times up to December 2019. All measurements took place approximately within the same period of the year (usually late October to December) in an effort to minimize seasonal phenomena. The Methana benchmark network was occupied using Leica receivers (SR9500 and AX1200). For each measurement period, the two benchmark stations were occupied for a period of 48 to 60 h, with time sampling of 15 s. Such time sampling was used for local-scale solutions with nearby continuous GNSS stations for which data were available with the same time sampling. While on regional-scale solutions, the data were resampled to 30-s data sets in order to be processed together with the freely distributed 30-s GNSS data available for European stations.

In early 2015, a continuous GNSS (cGNSS) station was also established in the Lygourio (LYGO) close to Methana peninsula as part of the HxGN SmartNet (https://gr.nrtk.eu/), that, ever since, has acted as local reference station. In addition, since August 2019, cGNSS data were available from a station (MTNA) located in the western part of Methana peninsula as part of the GNSS network of the Institute of Geodynamics, National Observatory of Athens (NOA). Horizontal velocity results from two more cGNSS stations were also available from the National Technical University of Athens (NTUA), i.e., station METH installed in 2004 in the southern sector of Methana peninsula, and from [55], i.e., station 010A located to the south-east of Methana peninsula, at Poros.

The available 30-s data from the two cGNSS stations (i.e., LYGO and MTNA) that were included in this study were processed separately on a daily basis together with several other cGNSS stations located in Greece as part of a regional network. From the time series of the station's coordinates, the cGNSS station velocity was estimated in the IGb08 reference frame (see Figures A1 and A2 in Appendix A).

The GPS data were processed using BERNESE software (v.5.2, University of Bern, Bern, Switzerland) [56]. The software allows for the estimation of epoch-wise receiver coordinates in Precise Point Positioning (PPP) mode as well as in the double difference mode. For the present research, the coordinates were calculated on a static mode. To get the best results, the PPP technique was used as a first step to get a priori coordinate values that were consequently introduced in the more precise double difference method. The GPS data from the benchmark stations and the other cGNSS stations in the broader area of Methana were processed together with several other EUropean REference Frame (EUREF) and cGNSS stations from Europe and Greece (http://www.epncb.oma.be) in order to define the local reference frame and to calculate the station coordinates. All measurement periods were processed with respect to the IGb08 reference frame (http://igscb.jpl.nasa.gov/network/refframe.html). Absolute antenna phase centre corrections were used in the data analysis. Several supporting files were introduced during the processing steps, aiming to define precise coordinates for the stations. Precise orbits from CODE analysis centre (ftp://ftp.unibe.ch/aiub/CODE) were used in the analysis, and a set of Earth orientation parameters was calculated. Crustal deformation caused by changing mass due to ocean tides and the atmosphere was taken into consideration, introducing auxiliary files. For the ocean loading effect, tide loading corrections were based on the FES2004 model (http://holt.oso.chalmers.se/loading). Atmospheric tidal loading coefficients were used from a global grid based on the Ray and Ponte (2003) model [57]. For troposphere modelling, the Vienna Mapping Function (http://ggosatm.hg.tuwien.ac.at) as well as the Neill mapping function were used. Ambiguities were solved using several resolution strategies based on the baseline length and the occupation time: (i) the wide-lane and narrow-lane techniques for the medium baselines (<10–200 km), where initially the L5 linear combination is processed (wide-lane) and introduced as known (fixed) in the subsequent run (narrow-lane) where the L3 linear combination is processed; (ii) the SIGMA strategy for the very short baselines (in between the two Methana stations, where both L1 and L2 data were available and used); and finally, (iii) the quasi-ionosphere-free (QIF) strategy for the long baselines (<1000–2000 km) were the main adopted ambiguity resolution strategies.

The long occupational time of the two GPS benchmark stations in Methana, during all measurement periods, resulted into low uncertainties (at a 95% confidence level) with overall errors of about 2.4–3.1 mm and 2.5–5.1 mm for the horizontal and vertical coordinates, respectively. The processing performed on daily sessions allowed for the extraction of a set of coordinates for each 24-h long period. The final set of coordinates was the mean value of all the sessions, and the errors led to quantification of the deviation of the daily coordinate solutions from the mean value.

3.3. Seismological Data and Processing

The digital seismological data used in this study were obtained from the database of the seismological laboratory of the National and Kapodistrian University of Athens (NKUA; http://dggsl.geol.uoa.gr/en_index.html).

The optimization of epicentres for the period 2006–2019 was carried out using the HYPOINVERSE software (v.1.4, United States Geological Survey: Menlo Park, CA, USA) [58]. The final quality of the relocated epicentres is very satisfactory due to the use of the one-dimension local velocity model developed by Karakonstantis et al. [59]. In particular, the root mean square (RMS) travel-time residual for the whole study period was determined to be 0.46 s, while the means of the horizontal and vertical direction uncertainties for the events' hypocentres are 1.51 km and 5.82 km, respectively.

The seismological data were divided into two time periods. The first one coincided with the time period that the GNSS data cover, i.e., January 2006 to August 2019, while the second period covered the time span of the interferometric processing, i.e., March 2015 to August 2019 (see Section 3.1).

4. Results and Discussion

4.1. Evidence from the GNSS Data Analysis

The observed velocities (i.e., V_E, V_N and V_U, along the east–west, north–south and vertical direction, respectively) and associated standard deviations (i.e., $STDV$) from all cGNSS and local benchmark stations are presented in Table 1. The LYGO cGNSS station exhibited SSE motion (Figure 2), calculated for a period of more than five years, that describes the anticipated regional motion of the area with respect to IGb08 [55] and is in good agreement with the two other cGNSS stations: METH and 010A. The newly established cGNSS station in the western Methana peninsula (i.e., MTNA) showed significantly different velocity due to the short period that the station was operating (i.e., from August to December 2019). The limited operational period reveals only a short portion of the station's annual trend and therefore could be biased by interannual variations. Consequently, these data have been excluded from further investigation but still briefly presented for future comparison (see Figure A1 in Appendix A). The velocity results from the two local benchmark stations in Methana peninsula (i.e., MENO and MESO) for a period of almost fourteen years (2006–2019) appeared consistent with the regional velocity field concerning the horizontal component (Figure 2). However, the southern station (i.e., MESO) revealed significant subsidence that has to be correlated with the MT-InSAR results (see Section 4.5).

Table 1. Velocity values of the continuous Global Navigation Satellite System (cGNSS) and local Global Positioning System (GPS) benchmark stations in Methana peninsula and the broader area in the IGb08 reference frame. Notation: V, velocity; $STDV$, standard deviation; E, east–west; N, north–south; and U, vertical/up.

Station	V_E (mm/year)	V_N (mm/year)	V_U (mm/year)	$STDV_E$ (mm/year)	$STDV_N$ (mm/year)	$STDV_U$ (mm/year)
LYGO	6.99	−12.97	−0.76	0.05	0.07	0.08
MTNA	7.02	3.35	4.88	1.26	1.76	6.37
MENO	6.50	−12.89	−0.21	0.21	0.33	0.30
MESO	7.02	−12.77	−5.24	0.15	0.08	1.12
METH *	6.77	−12.51	–	0.45	0.44	–
010A *	7.41	−13.01	–	0.43	0.44	–

* Values provided by the National Technical University of Athens (NTUA) and [55].

Aiming to better define the local deformation in Methana peninsula, the regional velocity vector (as it was estimated in LYGO cGNSS station) was subtracted from the two local GPS benchmark stations as well as the two other cGNSS sites (i.e., METH and 010A). Although the resulting horizontal velocity vectors are very small (<1 mm/year) with significant errors (Table 2) indicating a uniform deformation of the broader Methana area, there is a pattern of differential motion in the area (Figure 3). The northern station (i.e., MENO) exhibits a westward motion, while the southern sites (i.e., MESO and METH) showed a northward motion. However, the latter is quite arbitrary due to the small horizontal velocity values and the resulting errors. The more evident differential behaviour between these two stations is in the vertical component (Table 2), with the southern station (i.e., MESO) exhibiting noticeable subsidence, as it was mentioned previously.

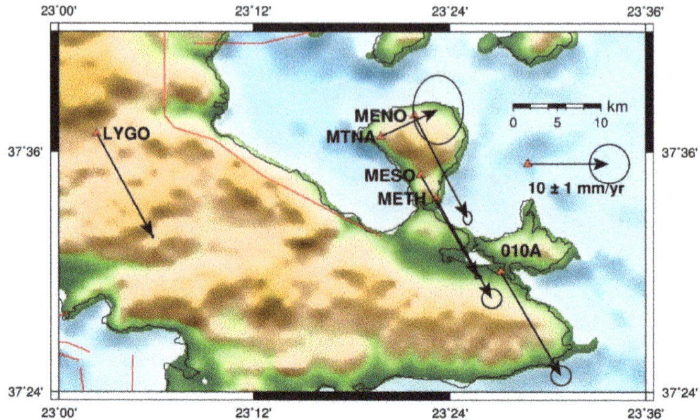

Figure 2. Horizontal velocity vectors from cGNSS and local GPS benchmark stations in the broader area of Methana peninsula with respect to the IGb08 reference frame. Stations METH and 010A were retrieved from other sources (NTUA and [55]).

Table 2. Local velocity values of the benchmark stations in Methana Peninsula with respect to LYGO station. Notation: *V*, velocity; *STDV*, standard deviation; E, east–west; N, north–south; and U, vertical/up.

Station	V_E (mm/year)	V_N (mm/year)	V_U (mm/year)	$STDV_E$ (mm/year)	$STDV_N$ (mm/year)	$STDV_U$ (mm/year)
LYGO	–	–	–			
MENO	−0.49	0.08	0.55	0.216	0.337	0.311
MESO	0.03	0.2	−4.48	0.158	0.106	1.123
METH *	−0.22	0.46	–	0.453	0.435	
010A *	0.42	−0.04	–	0.433	0.445	

* Values provided by NTUA and [55].

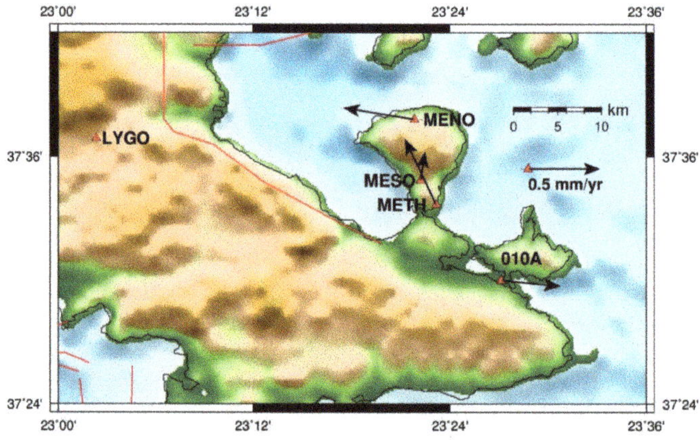

Figure 3. Local horizontal velocity field calculated with respect to the LYGO cGNSS site.

4.2. Observations from the Seismic Data Analysis

The analysis of the seismic data shows that the Saronikos Gulf region is mainly characterized by moderate seismic events and shallow depths, as already discussed in the literature [29]. During the period from 2006 to February 2015, the most important earthquakes occurred in 2008, 2012 and 2014, with magnitudes (Mw) of 3.4 to 4.2 and focal depths of 8 to 16 km. The focal mechanism solutions indicate the activation of NW-SE or NE-SW striking normal faults (Figure 4). Two clusters of intense seismic activity are located offshore: between Methana and Poros Island, and NE of Aegina Island. Shallow microseismicity is observed NW of Saronikos Gulf (in an area where the 2012 and Mw 4.1 earthquake occurred), offshore NW of Aegina island and at the southern coast of Attica (Figure 4).

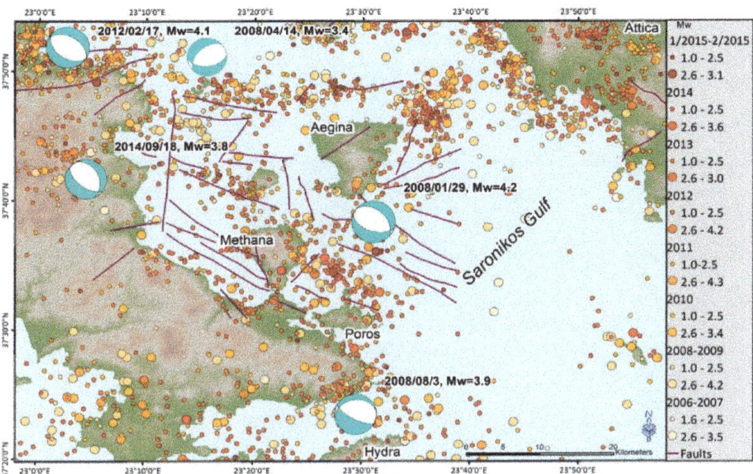

Figure 4. Seismic activity recorded in 2006–February 2015.

For the period 2015–2019, there were three significant seismic events in the area: two moderate shallow earthquakes with magnitudes Mw = 3.8 and Mw = 3.6 and a very deep one (depth = 142 km) with magnitude Mw = 4.2 (Figure 5). In this period, a significant seismic swarm was observed NE of Poros Island. The seismic activity in this cluster started in 2016 and continued with few events up to 2019. Additional to the surface seismicity, very deep (80–100 km) sporadic micro-events took place on a wider area extending from Methana towards Hydra Island. These deep micro-events may be attributed to the NE-subduction zone of the Ionian oceanic plate that reaches very deep in this area [5,54,59,60].

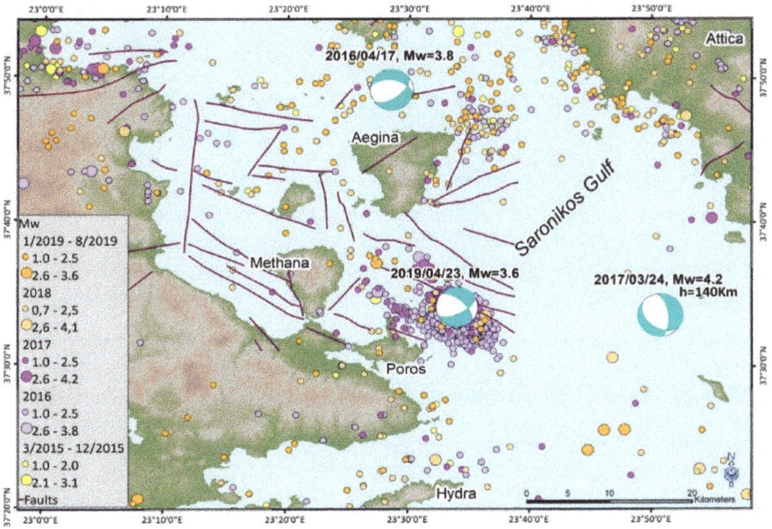

Figure 5. Seismic activity recorded in March 2015–August 2019.

4.3. PS and DS Distribution

After MT-InSAR processing, 4769 PS targets and 6234 DS targets were retrieved (Figure 6). The spatial densities of the targets in the area were calculated as 95.14 PS/km^2 and 124.36 DS/km^2 by considering the whole landmass of the peninsula (approximately 50 km^2). Most of the targets are located at low altitudes (up to approximately 100–150 m a.s.l. for PS and up to approximately 350–400 m a.s.l. for DS), where there are urban settlements, and also on geological formations which exhibit high coherence, such as volcanoclasts. PS targets have a coherence greater than 0.70, while the DS have coherence greater than 0.79.

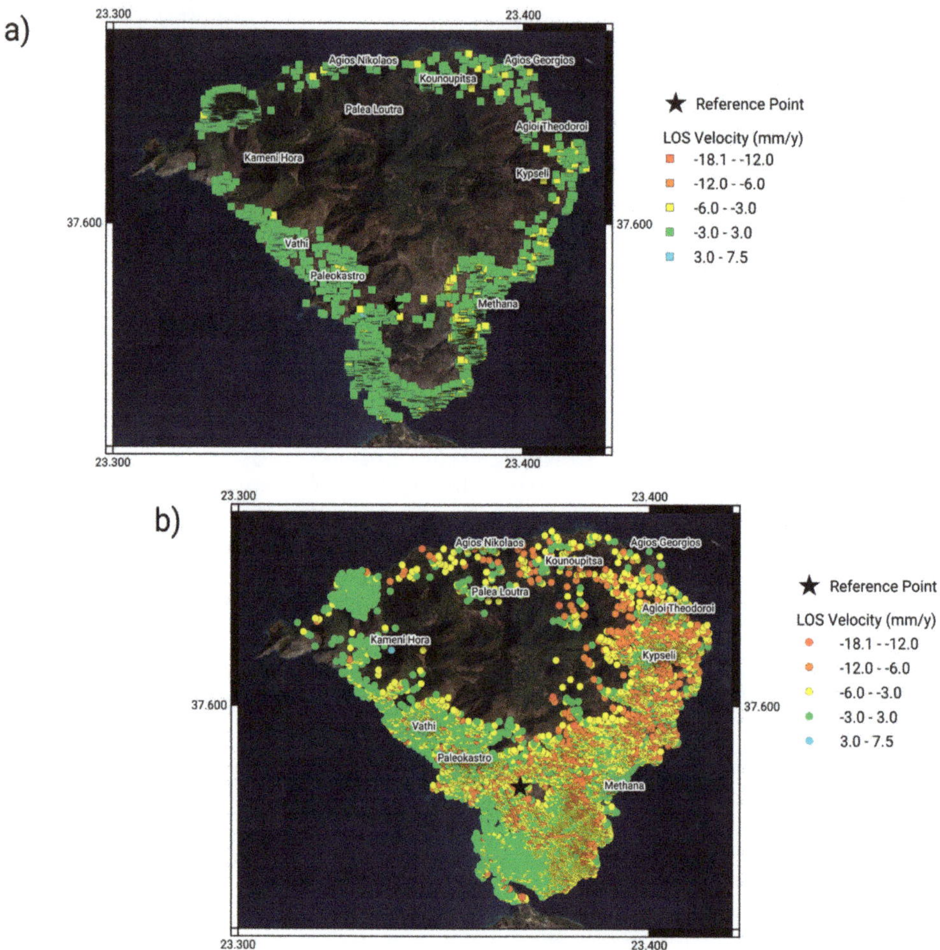

Figure 6. Line-of-sight (LOS) deformation velocity in 2015–2019 from the Multi-Temporal Interferometric Synthetic Aperture Radar (MT-InSAR) analysis for the (**a**) Persistent Scatterers (PS) and (**b**) Distributed Scatterers (DS) datasets, overlapped onto satellite optical imagery (© 2020 TerraMetrics, © 2020 Google).

Very dense coverage of PS is observed at the Mavri Petra andesitic dome north of Kameni Chora village due to the strong coherence over areas with sparse vegetation and bare ground, in particular over large blocks of agglomerate and cinder (Figure 7). The Mavri Petra unit has the greatest density of PS with 1241 PS/km^2 and a density of DS of 198 DS/km^2. The density of PS on Mavri Petra is even greater than in Methana town, which is an urban area (see Section 4.5). Methana town has 794 PS/km^2

and 479 DS/km^2. The lowest density of PS was observed in alluvial areas due to much lower coherence, with 156 PS/km^2 and 195 DS/km^2. Volcanoclastic units have densities of 397 PS/km^2 and 784 DS/km^2.

Figure 7. (**a**) Aerial view of the Mavri Petra andesitic dome and (**b**) details from the south-east, including Kameni Chora village (photographs © TobiasSchorr@Methana.de).

4.4. MT-InSAR Data Precision, Calibration and Validation

The present study is based on ascending SAR dataset only (see Section 3.1). Therefore, all results and values reported hereinafter are LOS values if not explicitly stated otherwise.

The uncertainty observed in the estimated V_{LOS} (i.e., standard deviation of the time series) is 2.8 mm/year on average for the PS dataset and 2.6 mm/year for the DS dataset. These values provide an indication of the precision of the MT-InSAR results and suggest that the ±3.0 mm/year V_{LOS} interval can be considered the velocity range, indicating stability.

On the other hand, in order to estimate the accuracy of the MT-InSAR results, these were validated using the local velocity values of the GPS benchmark stations MESO and MENO in Methana Peninsula with respect to LYGO cGNSS station (Table 2). A frame of reference transformation was applied to the velocity vectors in order to calculate the estimated LOS velocity of the two GPS benchmark stations and to allow for comparison with the MT-InSAR data.

Knowledge of the Euler angles (i.e., ω_x, ω_y and ω_z; Table 3) of the Sentinel-1 LOS along the ascending relative orbit 102 allowed us to generate the rotation table for each axis: R_1, R_2 and R_3 via Equations (1)–(3), respectively:

$$R_1 = \begin{bmatrix} 1 & 0 & 0 \\ 0 & \cos\omega_x & \sin\omega_x \\ 0 & -\sin\omega_x & \cos\omega_x \end{bmatrix} \tag{1}$$

$$R_2 = \begin{bmatrix} \cos\omega_y & 0 & -\sin\omega_y \\ 0 & 1 & 0 \\ \sin\omega_y & 0 & \cos\omega_y \end{bmatrix} \tag{2}$$

$$R_3 = \begin{bmatrix} \cos\omega_z & \sin\omega_z & 0 \\ -\sin\omega_z & \cos\omega_z & 0 \\ 0 & 0 & 1 \end{bmatrix} \tag{3}$$

Table 3. Euler angles of the Sentinel-1 LOS along ascending mode relative orbit 102.

Euler Angle	Degrees	Physical Interpretation
ω_x	0	–
ω_y	−37.1	Incidence angle
ω_z	11	Declination from the north (azimuth)

The right-hand rule was used to calculate the Euler angles because the reference frames are right-handed. The incidence angle for the area where the two GPS benchmark stations are found (namely, ω_y) was estimated using the European Space Agency (ESA)'s SeNtinel Application Platform (SNAP) software — Sentinel-1 toolbox S1TBX (v.7.0, SkyWatch, Waterloo, Canada),while the azimuth angle of the ascending relative orbit 102 in the area (namely, ω_z) was estimated using QGIS software (v.3.14, QGIS.ORG, Switzerland).

Then, using Equation (4), the rotation matrix (5) was composed and finally exploited to transform the reference frame into the LOS geometry:

$$R = R_1(\omega_x) \cdot R_2(\omega_y) \cdot R_3(\omega_z) \tag{4}$$

$$R = \begin{bmatrix} 0.78293 & 0.15219 & 0.60321 \\ -0.19081 & 0.98163 & 0 \\ -0.59213 & -0.11510 & 0.79758 \end{bmatrix} \tag{5}$$

The velocity vectors of GPS benchmark stations MENO, i.e., (6), and MESO, i.e., (7), were transformed to the new reference of frame using Equation (8):

$$V_{\text{MENO}} = \begin{bmatrix} -0.49 \\ 0.08 \\ 0.55 \end{bmatrix} \text{mm/year} \tag{6}$$

$$V_{\text{MESO}} = \begin{bmatrix} 0.03 \\ 0.20 \\ -4.48 \end{bmatrix} \text{mm/year} \tag{7}$$

$$V' = R \cdot V \tag{8}$$

The LOS velocity of the two benchmark stations was finally estimated: $V_{\text{LOS}} = -0.16$ mm/year for MENO and $V_{\text{LOS}} = -3.61$ mm/year for MESO.

We calibrated the PS and DS datasets by tying the estimated V_{LOS} at the closest scatterer to the MESO station (i.e., 40 m away) to the value measured at the GPS station (i.e., −3.61 mm/year).

As the scatterer originally exhibited a V_{LOS} of +0.26 mm/year, the adjustment meant a V_{LOS} shift of −3.87 mm/year, which was applied to all PS and DS targets.

At MENO station, the calibrated V_{LOS} from the MT-InSAR analysis equals −0.19 mm/year, thus matching extremely well with the −0.16 mm/year V_{LOS} measured at the GPS station and providing validating evidence for the MT-InSAR dataset.

To account for the velocity shift due to calibration of the PS and DS targets to MESO station, their time series were also adjusted as follows [61]:

$$d_i(t_n) = d_{i_{old}}(t_n) - V_{sh} * [t_n - t_0] \tag{9}$$

where $d_i(t_n)$ is the adjusted position of target i at time t_n, $d_{i_{old}}(t_n)$ is its original position at time t_n, V_{sh} is the V_{LOS} shift and t_0 is the first date of the time series.

4.5. Interpretation of the Observed Ground Deformation

The maximum V_{LOS} away from the sensor (i.e., negative values) that is observed after calibrations is −8.5 mm/year from the PS dataset and −18.1 mm/year from the DS dataset (Figure 6). PS and DS with the strongest movement away from the sensor are located on the eastern flank of the peninsula. Considering the ascending mode-looking geometry and the orientation of the LOS, this could plausibly indicate the occurrence of downslope mass movements. Compared with the topography displayed in Figure 1b, these movements concentrate on the volcanic ranges and gully erosion landforms above the urban settlements of Methana, Kipseli, Agii Teodori and Agios Georgios. Photo interpretation using Google Earth imagery, however, did not highlight changes of specific landforms for the 2012–2019 period.

The maximum V_{LOS} towards the sensor (i.e., positive values) from the PS dataset is +7.5 mm/year, and that from the DS dataset is +3.8 mm/year (Figure 6). Some values above +3.0 mm/year are observed for very few PS and DS scatterers, mainly located on the western flank of Methana peninsula. These scatterers are, however, characterized by a high uncertainty of their deformation rate (i.e., up to 4.0–5.0 mm/year). Their estimated velocities are therefore not considered as robust as those of other scatterers characterized by higher precision.

The northern portion of Methana town is built onto alluvium, and it seems more stable (V_{LOS} = −1.8 mm/year) than the central zone of the town, which is moving away from the satellite at −2.5 mm/year (Figure 8a), although both zones show velocities within the range of stability. The central zone is built onto volcanoclasts (Figure 8c), and the southern zone onto limestone (−2.9 mm/year). The southern zone is moving away from the satellite with relatively low V_{LOS}. Overall, it appears that the majority of the urban area footprint (Figure 8d) is not affected by significant deformation, while somehow, greater rates are found in the suburban zones, as revealed by the DS targets.

The deformation scenario at Vathi and Paleokastro (Figure 9a) shows that there is a slight movement away from the satellite also in this sector of the peninsula. Similar to what was observed for Methana town, PS targets in the region indicates that the urban areas have a mean value of −0.9 mm/year, while the DS targets for the suburban and more rural zones show −3.1 mm/year.

Annual LOS velocities do not reveal any significant deformation patterns across the sparsely vegetated outcrops at Mavri Petra (Figure 10a), which mostly shows both PS and DS with V_{LOS} in the range of ±3.0 mm/year and absence of any clear deformation patterns from the sole observation of the annual LOS velocities. However, evidence of seasonal deformation can be detected in the time series at several locations across the peninsula, including at Mavri Petra (Figure 10b) and at many other sites, for instance at Vathi and Paleokastro (Figure 9b) and at Methana town (Figure 8b). However, in the latter, this happens with less prominent amplitudes.

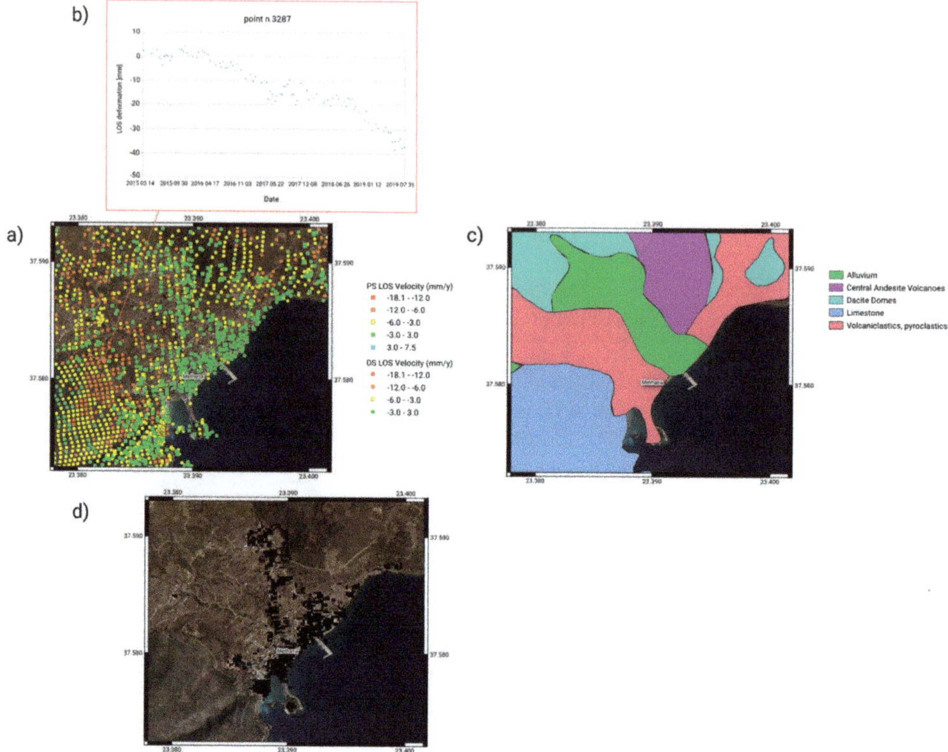

Figure 8. (**a**) LOS deformation velocity in 2015–2019 from the MT-InSAR analysis, (**b**) example of time series for a PS located on limestone, (**c**) geology/lithology and (**d**) urban footprint of Methana town: (**a,c,d**) are overlapped onto satellite optical imagery (© 2020 TerraMetrics, © 2020 Google).

The detected seasonal fluctuations at Mavri Petra are significant, both for PS with LOS velocity close to 0 mm/year and for PS with higher LOS velocity. Those displacements have an amplitude of approximately 5 mm and periodicity of 1 year (Figure 10b). Similar amplitudes are found at Vathi and Paleokastro (Figure 9b). The peaks of the fluctuations generally occur in the period October–March, i.e., rainfall season [62]. This suggests the presence of a seasonal component in the deformation behaviour that is potentially influenced by rainfall and overlaps onto the long-term, generally linear, trend observed across the peninsula in 2015–2019. In this regard, seasonal signals are commonly observed in surface deformation records in other sites worldwide (e.g., [63]), with annual or even semi-annual periods and different amplitudes from negligible to significant, depending on their driving causes. Future ground deformation investigations in Methana could be focused on identification and removal of such periodical terms via advanced time series processing. This operation would better enhance the long-term deformation behaviour of the observed sites which, in this case, may be partly concealed by these fluctuations.

Higher LOS velocities (e.g., −5.0 to −10.0 mm/year) have been observed along steep valleys, mainly on slopes with east orientation (Figure 11a), where visibility of the ascending mode LOS eases the detection of downslope mass movements. The observed displacement could therefore be related to extremely to very slow mass movements and slope instability affecting rugged terrain and volcanic landforms that are recognized as prone to landslides and rock-falls [18]. Some seasonal fluctuations are also detected in the time series (Figure 11b).

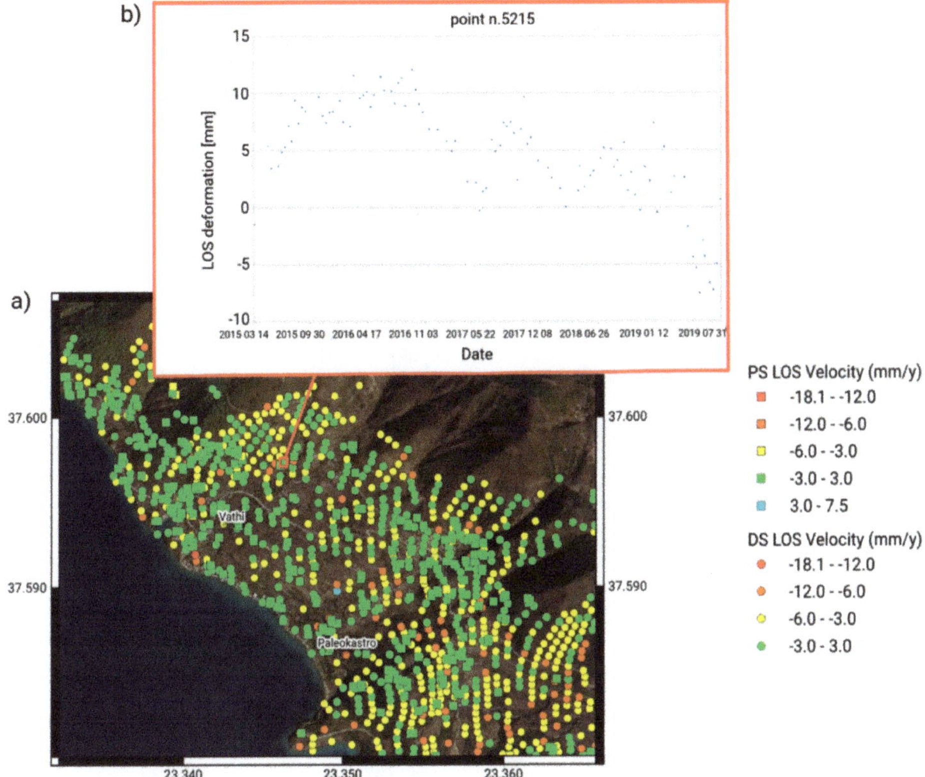

Figure 9. (**a**) LOS deformation velocity over Vathi and Paleokastro in 2015–2019 from the MT-InSAR analysis overlapped onto satellite optical imagery (© 2020 TerraMetrics, © 2020 Google) and (**b**) example of time series for a PS located near Vathi town.

When the evidence from the MT-InSAR investigation is integrated with the outcomes of the GNSS data analysis (see Section 4.1), it can be suggested that the ground motions that are observed from 2015 to 2019 are compatible with the anticipated "low" volcanic activity of Methana. This comes out very evidently from the PS dataset and the generally low V_{LOS}. Despite the low point density and spatial coverage of the volcanic edifice, the overall spatial distribution of the PS deformation values does not reveal a deformation field that could be reliably attributed to a typical volcanic inflation/deflation dynamic of the whole Methana, as found in the literature for other active volcanic areas. On the other hand, the DS dataset may suggest the presence of more widespread deformation patterns along the eastern flank. However, there are only few cases where the observed motions associate with specific landforms (e.g., narrow valleys, erosion gullies and superficial slides). In this regard, the photo interpretation of the slopes through Google Earth optical imagery did not highlight specific circumstances worth further investigation.

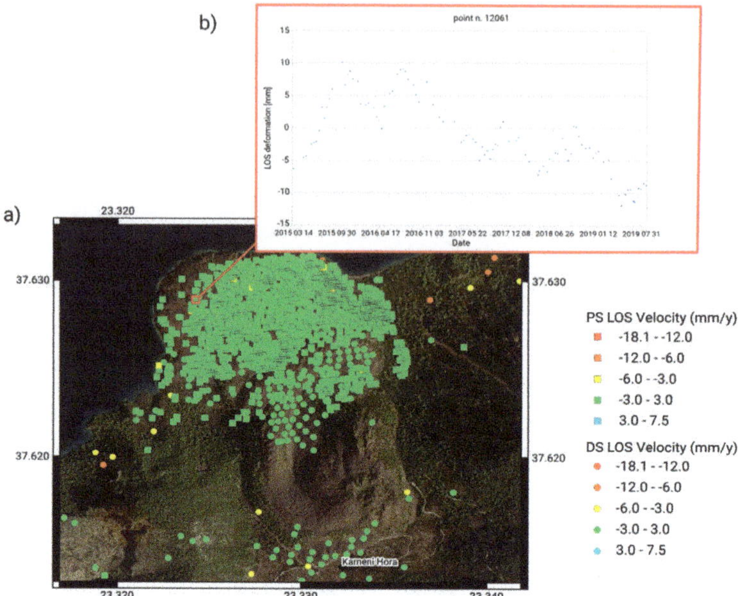

Figure 10. (**a**) LOS deformation velocity over Mavri Petra in 2015–2019 from the MT-InSAR analysis overlapped onto satellite optical imagery (© 2020 TerraMetrics, © 2020 Google) and (**b**) example of time series for a PS located in the NW portion of Mavri Petra.

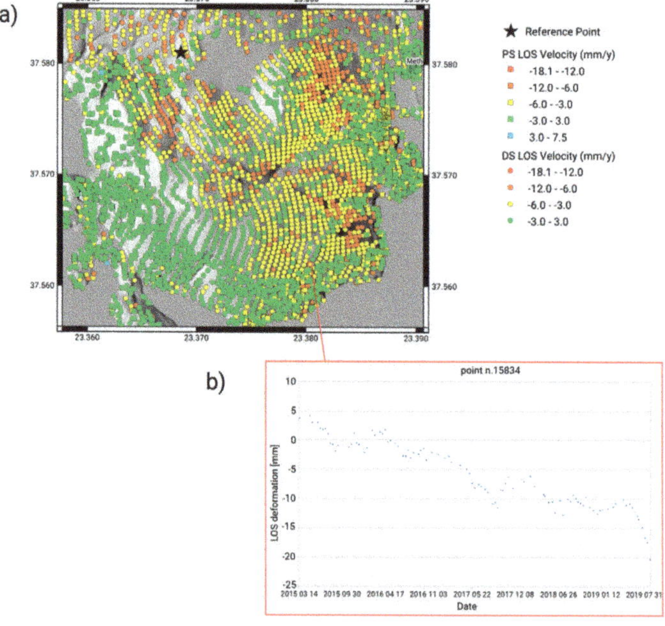

Figure 11. (**a**) LOS deformation velocity over the southern portion of Methana peninsula in 2015–2019 from the MT-InSAR analysis overlapped onto shaded relief of the 5.5-m digital surface model by the Hellenic Mapping and Cadastral Organisation and (**b**) example of time series for a DS located in the SE part of Methana peninsula.

Due to the lack of strong seismic events (Mw > 4) or intense seismicity in the area of Methana peninsula (see Section 4.2), the observed ground deformation could not be associated with the recorded seismicity. There are active seismic zones in the vicinity of the peninsula; however, neither the location, the depth nor the magnitude of these events could be directly associated with the observed motions. Even if a seismic contribution to triggering superficial mass movements was present, this would be plausibly limited to nearby seismic swarms. However, there is no hard evidence from ground observations to support further such a hypothesis.

Accounting for the geothermal activity and warm springs at various locations in Methana, the only situation where motions were found close to a thermal spring is in the immediate vicinity of the open-air thermal baths of the Thermal Spa in the southern part of Methana town (see Figure 8a). However, the LOS deformation rate values found in the PS dataset most likely associate with rock-fall accumulation overlooking the road (Figure 12), while those in the DS dataset cannot be easily separated from the wider deformation pattern extending across the above topography (see Figure 11). The seasonality observed in the MT-InSAR time series in this area as well as at many other locations of the peninsula appear to be more plausibly due to soil moisture and groundwater level oscillations, partly controlled by precipitation, as discussed above. On the other hand, another process that could explain such seasonality could be linked with fluctuations in the geothermal reservoir located in the centre of Methana at a depth of approximately 2 km [64], though more investigation would be needed to verify the potential occurrence of such a process.

Figure 12. (a) LOS deformation velocity of the PS over Methana town in 2015–2019 from the MT-InSAR analysis overlapped onto satellite optical imagery (© 2020 TerraMetrics, © 2020 Google) and (b,c) ground-truth photographs of rock-fall accumulation along the Methana–Taktikopolis road (© 2020 Google).

5. Conclusions

In the context of the abundant literature of satellite InSAR studies on the south Aegean volcanic arc and the published geological and geochemical research on Methana, the present study is the first that undertakes a multi-temporal InSAR analysis of the most recent ground motions in the peninsula in the period 2015–2019 and attempts to find correlations with local GPS benchmark, cGNSS and regional seismicity data collected since 2006. Methana is located in the seismically active neotectonic basin constituting the Saronic Gulf area, and recent efforts have been made by Greek institutions to specifically monitor the peninsula, among which is the nomination of the Permanent Scientific Committee "Greek Volcano Arc Monitoring" by the administrative board of the Earthquake Planning and Protection Organization (EPPO) in June 2020. In this context, this study is timely in starting to put together initial geodetic and satellite datasets to create a virtual monitoring network that, in the future, could be developed into a proper volcano observatory system.

Based on the integrated data analysis, the general conclusion of this research is that the ground motions observed in the Sentinel-1 MT-InSAR ascending mode data are compatible with the anticipated "low" volcanic activity of Methana. Although the volcano does not show clear signs of activity in recent years, it is a hotspot of potential concern given that it is adjacent to very high-density population areas, including Athens and the surrounding major towns. The interpreted low level of activity is clearly the initial baseline above which further investigations need to be conducted and will need to be verified once the whole archive of SAR datasets dating back to the early 1990s is analysed (e.g., ERS-1/2 and ENVISAT) and the ascending geometry of Sentinel-1 made available for this research is complemented with same InSAR analysis of the matching descending dataset acquired over the peninsula since 2014. It is evident that the use of one SAR geometry only limits the extent to which the observed LOS ground motions can be interpreted. Therefore, future research may include processing of SAR data (even at higher spatial resolution) from both ascending and descending geometries and reconstruction of the 3D deformation field as well as longer time series of geodetic data to improve our understanding of the long-term deformation behaviour of the peninsula.

Author Contributions: Conceptualization, F.C., D.T. and I.P.; methodology, T.G., F.C., D.T. and V.S.; software, T.G., V.S. and K.P.; validation, T.G., V.S. and K.P.; formal analysis, T.G., F.C., D.T., V.S. and K.P.; investigation, T.G., F.C., D.T., V.S. and K.P.; resources, I.P.; data curation, T.G., V.S. and K.P.; writing—original draft preparation, T.G., F.C., D.T., V.S. and K.P.; writing—review and editing, I.P.; supervision, F.C. and I.P. All authors have read and agreed to the published version of the manuscript.

Funding: Remeasurement of the GPS benchmarks MESO and MENO was financed by "HELPOS—Hellenic Plate Observing System" (MIS 5002697), which is implemented under the action "Reinforcement of the Research and Innovation Infrastructure", funded by the Operational Programme "Competitiveness, Entrepreneurship and Innovation" (NSRF 2014-2020) and co-financed by Greece and the European Union (European Regional Development Fund).

Acknowledgments: Copernicus Sentinel-1 SAR data were accessed through the Sentinel Hub platform (https://scihub.copernicus.eu/) and processed using the GAMMA SAR and Interferometry software licensed to Harokopio University of Athens and ESA's SNAP software. GNSS data were provided by the National Technical University of Athens (NTUA) for METH station, by HxGN SmartNet for LYGO station, and by the Institute of Geodynamics of the National Observatory of Athens (NOA) for MTNA and 010A stations. Seismic data were provided by the Department of Geophysics and Geothermy, National and Kapodistrian University of Athens (NKUA) (http://dggsl.geol.uoa.gr/en_index.html). High-resolution DSM was provided by the Hellenic Mapping and Cadastral Organisation. Some of the figures were created with the Generic Mapping Tool (GMT) software [65].

Conflicts of Interest: The authors declare no conflict of interest.

Appendix A

Daily solutions for the continuous GNSS stations in Lygourio (LYGO) and in the western part of Methana (MTNA) are provided in Figures A1 and A2, respectively.

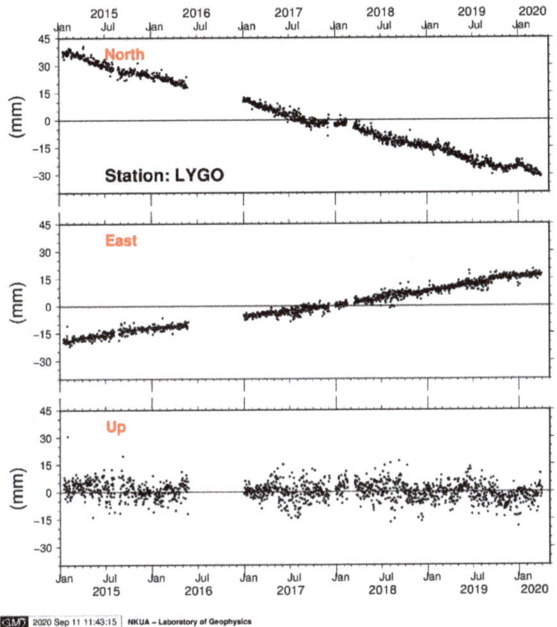

Figure A1. Time series for the three components of the daily solutions, spanning from January 2015 to March 2020, for the continuous GNSS station in the Lygourio (LYGO) area, west of Methana peninsula.

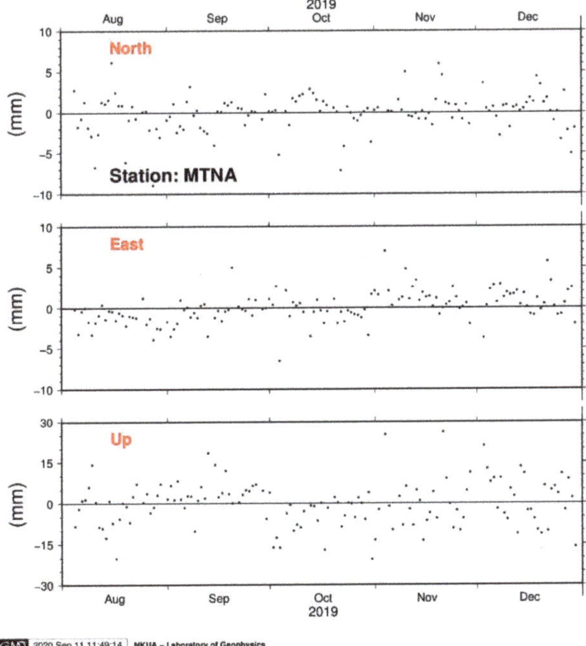

Figure A2. Time series for the three components of the daily solutions, spanning from August to December 2020, for the continuous GNSS station MTNA, located in the western part of Methana peninsula.

References

1. Zouros, N. Assessment, protection, and promotion of geomorphological and geological sites in the Aegean area, Greece. *Géomorphosites Définition Évaluation Cartogr.* **2005**, *11*, 227–234. [CrossRef]
2. Kyriakopoulos, K.G. Volcanoes "Monuments of Nature". In *Natural Heritage from East to West: Case Studies from 6 EU Countries*; Springer: Berlin/Heidelberg, Germany, 2010; pp. 59–70. ISBN 9783642015762.
3. D'Alessandro, W.; Brusca, L.; Kyriakopoulos, K.; Michas, G.; Papadakis, G. Methana, the westernmost active volcanic system of the south Aegean arc (Greece): Insight from fluids geochemistry. *J. Volcanol. Geotherm. Res.* **2008**, *178*, 818–828. [CrossRef]
4. Vougioukalakis, G.E.; Fytikas, M. Volcanic hazards in the Aegean area, relative risk evaluation, monitoring and present state of the active volcanic centers. *Dev. Volcanol.* **2005**, *7*, 161–183. [CrossRef]
5. Makris, J.; Papoulia, J.; Drakatos, G. Tectonic deformation and microseismicity of the Saronikos Gulf, Greece. *Bull. Seismol. Soc. Am.* **2004**, *94*, 920–929. [CrossRef]
6. Scientists Decide to Increase Monitoring of Methana Volcano Near Athens|Neos Kosmos. Available online: https://neoskosmos.com/en/130406/scientists-decide-increase-monitoring-of-methana-volcano-near-athens/ (accessed on 14 August 2020).
7. NOANET, Station Mtna. Available online: http://geodesy.gein.noa.gr:8000/nginfo/mtna/ (accessed on 14 August 2020).
8. Parcharidis, I.; Lagios, E. Deformation in Nisyros volcano (Greece) using differential radar interferometry. *Bull. Geol. Soc. Greece* **2001**, *34*, 1587. [CrossRef]
9. Ganas, A.; Lagios, E.; Dietrich, V.J.; Vassilopoulou, S.; Hurni, L.; Stavrakakis, G. Interferometric Mapping of Ground Deformation in Nissyros Volcano, Aegean Sea, during 1995-1996. In Proceedings of the 6th Pan-Hellenic Geographical Congress, Thessaloniki, Greece, 3–6 October 2002; Volume II, pp. 135–141.
10. Sachpazi, M.; Kontoes, C.; Voulgaris, N.; Laigle, M.; Vougioukalakis, G.; Sikioti, O.; Stavrakakis, G.; Baskoutas, J.; Kalogeras, J.; Lepine, J.C. Seismological and SAR signature of unrest at Nisyros caldera, Greece. *J. Volcanol. Geotherm. Res.* **2002**, *116*, 19–33. [CrossRef]
11. Parcharidis, I.; Sakkas, V.A.; Ganas, A.; Katopodi, A. A production of Digital Elevation Model (DEM) by means of SAR tandem images in a volcanic landscape and its quality assessment. In Proceedings of the 6th Pan-Hellenic Geographical Conference of the Hellenic Geographical Society, Thessaloniki, Greece, 3–6 October 2002; Volume 2, pp. 201–207.
12. Sykioti, O.; Kontoes, C.C.; Elias, P.; Briole, P.; Sachpazi, M.; Paradissis, D.; Kotsis, I. Ground deformation at Nisyros volcano (Greece) detected by ERS-2 SAR differential interferometry. *Int. J. Remote Sens.* **2003**, *24*, 183–188. [CrossRef]
13. Parcharidis, I.; Lagios, E.; Sakkas, V. Differential interferometry as a tool of an early warning system in reducing the volcano risk: The case of Nisyros Volcano. In Proceedings of the of the 10th Congress of the Geological Society of Greece, Athens, Greece, 15–17 April 2004; Extended Abstracts. National Documentation Centre: Athina, Greece, 2004; pp. 913–918.
14. Lagios, E.; Sakkas, V.; Parcharidis, I.; Dietrich, V. Ground deformation of Nisyros Volcano (Greece) for the period 1995–2002: Results from DInSAR and DGPS observations. *Bull. Volcanol.* **2005**, *68*, 201–214. [CrossRef]
15. Mouratidis, A.F.; Tsakiri-Strati, M.; Astaras, T. SAR Interferometry for Geoscience applications in Greece. In *The Apple of Knowledge: Volume in Honor of Prof. D. Arabelos*; Kaltsikis, C., Kontadakis, M.E., Spatalas, S., Tziavos, I.N., Tokmakidis, K., Eds.; Publication of the School of Rural & Surveying Engineering: Thessaloniki, Greece, 2010; pp. 340–350.
16. Foumelis, M.; Trasatti, E.; Papageorgiou, E.; Stramondo, S.; Parcharidis, I. Monitoring Santorini volcano (Greece) breathing from space. *Geophys. J. Int.* **2013**, *193*, 161–170. [CrossRef]
17. Papageorgiou, E.; Foumelis, M.; Trasatti, E.; Ventura, G.; Raucoules, D.; Mouratidis, A. Multi-sensor SAR geodetic imaging and modelling of santorini volcano post-unrest response. *Remote Sens.* **2019**, *11*, 259. [CrossRef]
18. Antoniou, V.; Nomikou, P.; Bardouli, P.; Lampridou, D.; Ioannou, T.; Kalisperakis, I.; Stentoumis, C.; Whitworth, M.; Krokos, M.; Ragia, L. An Interactive Story Map for the Methana Volcanic Peninsula. In Proceedings of the 4th International Conference on Geographical Information Systems Theory, Applications and Management (GISTAM 2018), Madeira, Portugal, 17–19 March 2018; Scitepress: Setúbal, Portugal, 2018; pp. 68–78.

19. Nomikou, P.; Papanikolaou, D.; Alexandri, M.; Sakellariou, D.; Rousakis, G. Submarine volcanoes along the aegean volcanic arc. *Tectonophysics* **2013**, *597–598*, 123–146. [CrossRef]

20. Francalanci, L.; Vougioukalakis, G.E.; Perini, G.; Manetti, P. A West-East Traverse along the magmatism of the south Aegean volcanic arc in the light of volcanological, chemical and isotope data. *Dev. Volcanol.* **2005**, *7*, 65–111. [CrossRef]

21. Pe-Piper, G.; Piper, D.J.W. The effect of changing regional tectonics on an arc volcano: Methana, Greece. *J. Volcanol. Geotherm. Res.* **2013**, *260*, 146–163. [CrossRef]

22. Popa, R.G.; Dietrich, V.J.; Bachmann, O. Effusive-explosive transitions of water-undersaturated magmas. The case study of Methana Volcano, South Aegean Arc. *J. Volcanol. Geotherm. Res.* **2020**, *399*, 106884. [CrossRef]

23. Dotsika, E.; Poutoukis, D.; Raco, B. Fluid geochemistry of the Methana Peninsula and Loutraki geothermal area, Greece. *J. Geochem. Explor.* **2010**, *104*, 97–104. [CrossRef]

24. Pe-Piper, G.; Piper, D.J.W. *The Igneous Rocks of Greece: The Anatomy of an Orogen*; Gebr. Borntraeger: Stuttgart, Germany, 2002; ISBN 9783443110307.

25. Papanikolaou, D.; Chronis, G.; Lykousis, V.; Pavlakis, P. *Submarine Neotectonic Map of Saronikos Gulf*; Earthquake, Planning and Protection Organization of Greece: Thessaloniki, Greece, 1989.

26. Pavlakis, P.; Lykoussis, V.; Papanikolaou, D.; Chronis, G. Discovery of a new submarine volcano in the western Saronic Gulf: The Paphsanias Volcano. *Bull. Hell. Geol. Soc. Greece* **1990**, *XXIV*, 59–70.

27. Fytikas, M.; Innocenti, F.; Kolios, N. The Plio-Quaternary volcanism of Saronikos area (western part of the active part of the Aegean volcanic arc). *Ann. Géologiques Pays Helléniques* **1988**, *33*, 23–45.

28. Armijo, R.; Meyer, B.; King, G.C.P.; Rigo, A.; Papanastassiou, D. Quaternary evolution of the Corinth Rift and its implications for the Late Cenozoic evolution of the Aegean. *Geophys. J. Int.* **1996**, *126*, 11–53. [CrossRef]

29. Makropoulos, K.; Kaviris, G.; Kouskouna, V. An updated and extended earthquake catalogue for Greece and adjacent areas since 1900. *Nat. Hazards Earth Syst. Sci.* **2012**, *12*, 1425–1430. [CrossRef]

30. Dietrich, V.J.; Mercolli, I.; Oberhänsli, R. Dazite, High-Alumina-Basalte und Andesite als Produkte amphiboldominierter Differentation (Aegina und Methan, Ägäischer Inselbogen). *Schweiz. Mineral. Petrogr. Mitt.* **1988**, *68*, 21–39.

31. Gaitanakis, P.; Dietrich, A.D. *Geological Map of Methana Peninsula, 1:25 000*; Swiss Federal Institute of Technology: Zurich, Switzerland, 1995.

32. Pe-Piper, G.; Piper, D.J.W. The South Aegean active volcanic arc: Relationships between magmatism and tectonics. *Dev. Volcanol.* **2005**, *7*, 113–133. [CrossRef]

33. Volti, T.K. Magnetotelluric measurements on the Methana Peninsula (Greece): Modelling and interpretation. *Tectonophysics* **1999**, *301*, 111–132. [CrossRef]

34. Curlander, J.C.; Mcdonough, R.N. *Synthetic Aperture Radar: Systems and Signal Processing*; Wiley: Hoboken, NJ, USA, 1991; ISBN 978-0-471-85770-9.

35. Torres, R.; Snoeij, P.; Geudtner, D.; Bibby, D.; Davidson, M.; Attema, E.; Potin, P.; Rommen, B.Ö.; Floury, N.; Brown, M.; et al. GMES Sentinel-1 mission. *Remote Sens. Environ.* **2012**, *120*, 9–24. [CrossRef]

36. Berger, M.; Moreno, J.; Johannessen, J.A.; Levelt, P.F.; Hanssen, R.F. ESA's sentinel missions in support of Earth system science. *Remote Sens. Environ.* **2012**, *120*, 84–90. [CrossRef]

37. Bamler, R.; Hartl, P. Synthetic aperture radar interferometry. *Inverse Probl.* **1998**, *14*, R1. [CrossRef]

38. Rosen, P.A.; Hensley, S.; Joughin, I.R.; Li, F.K.; Madsen, S.N.; Rodriguez, E.; Goldstein, R.M. Synthetic aperture radar interferometry. *Proc. IEEE* **2000**, *88*, 333–380. [CrossRef]

39. Crosetto, M.; Monserrat, O.; Cuevas-González, M.; Devanthéry, N.; Crippa, B. Persistent Scatterer Interferometry: A review. *ISPRS J. Photogramm. Remote Sens.* **2016**, *115*, 78–89. [CrossRef]

40. Amelung, F.; Galloway, D.L.; Bell, J.W.; Zebker, H.A.; Laczniak, R.J. Sensing the ups and downs of Las Vegas: InSAR reveals structural control of land subsidence and aquifer-system deformation. *Geology* **1999**, *27*, 483–486. [CrossRef]

41. Dixon, T.H.; Amelung, F.; Ferretti, A.; Novali, F.; Rocca, F.; Dokka, R.; Sella, G.; Kim, S.W.; Wdowinski, S.; Whitman, D. Space geodesy: Subsidence and flooding in New Orleans. *Nature* **2006**, *441*, 587–588. [CrossRef]

42. Herrera, G.; Davalillo, J.C.; Mulas, J.; Cooksley, G.; Monserrat, O.; Pancioli, V. Mapping and monitoring geomorphological processes in mountainous areas using PSI data: Central Pyrenees case study. *Nat. Hazards Earth Syst. Sci.* **2009**, *9*, 1587–1598. [CrossRef]

43. Cigna, F.; Tapete, D.; Garduño-Monroy, V.H.; Muñiz-Jauregui, J.A.; García-Hernández, O.H.; Jiménez-Haro, A. Wide-area InSAR survey of surface deformation in urban areas and geothermal fields in the eastern Trans-Mexican Volcanic Belt, Mexico. *Remote Sens.* **2019**, *11*, 2341. [CrossRef]

44. Spaans, K.; Hooper, A. InSAR processing for volcano monitoring and other near-real time applications. *J. Geophys. Res. Solid Earth* **2016**, *121*, 2947–2960. [CrossRef]

45. Cigna, F.; Bianchini, S.; Casagli, N. How to assess landslide activity and intensity with Persistent Scatterer Interferometry (PSI): The PSI-based matrix approach. *Landslides* **2013**, *10*, 267–283. [CrossRef]

46. Ferretti, A.; Prati, C.; Rocca, F. Nonlinear subsidence rate estimation using permanent scatterers in differential SAR interferometry. *IEEE Trans. Geosci. Remote Sens.* **2000**, *38*, 2202–2212. [CrossRef]

47. Ferretti, A.; Prati, C.; Rocca, F. Permanent scatterers in SAR interferometry. *IEEE Trans. Geosci. Remote Sens.* **2001**, *39*, 8–20. [CrossRef]

48. Berardino, P.; Fornaro, G.; Lanari, R.; Sansosti, E. A new algorithm for surface deformation monitoring based on small baseline differential SAR interferograms. *IEEE Trans. Geosci. Remote Sens.* **2002**, *40*, 2375–2383. [CrossRef]

49. Ferretti, A.; Fumagalli, A.; Novali, F.; Prati, C.; Rocca, F.; Rucci, A. A new algorithm for processing interferometric data-stacks: SqueeSAR. *IEEE Trans. Geosci. Remote Sens.* **2011**, *49*, 3460–3470. [CrossRef]

50. Shamshiri, R.; Nahavandchi, H.; Motagh, M.; Hooper, A. Efficient Ground Surface Displacement Monitoring Using Sentinel-1 Data: Integrating Distributed Scatterers (DS) Identified Using Two-Sample t-Test with Persistent Scatterers (PS). *Remote Sens.* **2018**, *10*, 794. [CrossRef]

51. Cigna, F.; Sowter, A. The relationship between intermittent coherence and precision of ISBAS InSAR ground motion velocities: ERS-1/2 case studies in the UK. *Remote Sens. Environ.* **2017**, *202*, 177–198. [CrossRef]

52. Werner, C.; Wegmüller, U.; Strozzi, T.; Wiesmann, A. Interferometric Point Target Analysis for Deformation Mapping. In Proceedings of the International Geoscience and Remote Sensing Symposium (IGARSS), Toulouse, France, 21–25 July 2003; Volume 7, pp. 4362–4364.

53. Farr, T.G.; Rosen, P.A.; Caro, E.; Crippen, R.; Duren, R.; Hensley, S.; Kobrick, M.; Paller, M.; Rodriguez, E.; Roth, L.; et al. The shuttle radar topography mission. *Rev. Geophys.* **2007**, *45*, RG2004. [CrossRef]

54. Papageorgiou, E. Crustal movements along the NW Hellenic volcanic arc from DGPS measurements. *Bull. Geol. Soc. Greece* **2010**, *43*, 331. [CrossRef]

55. Chousianitis, K.; Ganas, A.; Gianniou, M. Kinematic interpretation of present-day crustal deformation in central Greece from continuous GPS measurements. *J. Geodyn.* **2013**, *71*, 1–13. [CrossRef]

56. Dach, R.; Lutz, S.; Walser, P.; Fridez, P. *Bernese GNSS Software Version 5.2*; User Manual; Astronomical Institute, University of Bern, Bern Open Publishing: Bern, Switzerland, 2015.

57. Ray, R.D.; Ponte, R.M. Barometric tides from ECMWF operational analyses. *Ann. Geophys.* **2003**, *21*, 1897–1910. [CrossRef]

58. Klein, F.W. *User's Guide to HYPOINVERSE, a Program for VAX Computers to Solve for Earthquake Locations and Magnitudes*; U.S. Geological Survey: Menlo Park, CA, USA, 1989.

59. Karakonstantis, A.; Papadimitriou, P.; Millas, C.; Spingos, I.; Fountoulakis, I.; Kaviris, G. Tomographic imaging of the NW edge of the Hellenic volcanic arc. *J. Seismol.* **2019**, *23*, 995–1016. [CrossRef]

60. Papanikolaou, D.; Lykousis, V.; Chronis, G.; Pavlakis, P. A comparative study of neotectonic basins across the Hellenic arc: The Messiniakos, Argolikos, Saronikos and Southern Evoikos Gulfs. *Basin Res.* **1988**, *1*, 167–176. [CrossRef]

61. Bateson, L.; Cuevas, M.; Crosetto, M.; Cigna, F.; Schijf, M.; Evans, H. *PANGEO: Enabling Access to Geological Information in Support of GMES: Deliverable 3.5 Production Manual. Version 1*; European Commission: Brussels, Belgium, 2012.

62. Climate-Data.org Methana Climate: Average Temperature, Weather by Month, Methana Water Temperature—Climate-Data.org. Available online: https://en.climate-data.org/europe/greece/methana/methana-27566/ (accessed on 18 August 2020).

63. Berti, M.; Corsini, A.; Franceschini, S.; Iannacone, J.P. Automated classification of Persistent Scatterers Interferometry time series. *Nat. Hazards Earth Syst. Sci.* **2013**, *13*, 1945–1958. [CrossRef]
64. Tzanis, A.; Efstathiou, A.; Chailas, S.; Lagios, E.; Stamatakis, M. The Methana Volcano—Geothermal Resource, Greece, and its relationship to regional tectonics. *J. Volcanol. Geotherm. Res.* **2020**, *404*, 107035. [CrossRef]
65. Wessel, P.; Smith, W.H.F. New, improved version of Generic Mapping Tools released. *EOS Trans AGU* **1998**, *79*, 579. [CrossRef]

 *applied sciences*

Article

The February-March 2019 Seismic Swarm Offshore North Lefkada Island, Greece: Microseismicity Analysis and Geodynamic Implications

Anastasios Kostoglou, Vasileios Karakostas, Polyzois Bountzis and Eleftheria Papadimitriou *

Geophysics Department, Aristotle University of Thessaloniki, 54124 Thessaloniki, Greece;
akostogl@geo.auth.gr (A.K.); vkarak@geo.auth.gr (V.K.); pmpountzp@geo.auth.gr (P.B.)
* Correspondence: ritsa@geo.auth.gr

Received: 24 May 2020; Accepted: 26 June 2020; Published: 29 June 2020

Featured Application: Authors are encouraged to provide a concise description of the specific application or a potential application of the work. This section is not mandatory.

Abstract: A quite energetic seismic excitation consisting of one main and three additional distinctive earthquake clusters that occurred in the transition area between the Kefalonia Transform Fault Zone (KTFZ) and the continental collision between the Adriatic and Aegean microplates is thoroughly studied after the high-precision aftershocks' relocation. The activated fault segments are in an area where historical and instrumental data have never claimed the occurrence of a catastrophic (M $\geq$ 6.0) earthquake. The relocated seismicity initially defines an activated structure extending from the northern segment of the Lefkada branch of KTFZ with the same NNE–SSW orientation and dextral strike slip faulting, and then keeping the same sense of motion, its strike becomes NE–SW and its dip direction NW. This provides unprecedented information on the link between the KTFZ and the collision front and sheds more light on the regional geodynamics. The earthquake catalog, which was especially compiled for this study, starts one year before the occurrence of the M_w5.4 main shock, and adequately provides the proper data source for investigating the temporal variation in the b value, which might be used for discriminating foreshock and aftershock behavior.

Keywords: seismic swarm; relocated aftershocks; transition zone; b value temporal variation; central Ionian Islands (Greece)

---

1. Introduction

A moderate magnitude M_w5.4 earthquake occurred on 5 February 2019 in the offshore area north of Lefkada Island, strongly felt in the Lefkada city and the onshore continental area to the east, with no major damage or injuries reported. In the area, an adequate number of moderate earthquakes have occurred in recent decades with not one having a magnitude larger than M6.0. The seismic excitation started with an M_w4.2 shock that occurred on 15 January 2019 to the northeast of the main shock which was followed by a very productive aftershock sequence with more than 250 located shocks by 15 March 2019.

The activated area is located at the boundary between the Kefalonia Transform Fault Zone (KTFZ) to the south and the Adriatic–Eurasian collision to the north. The continental collision is expressed by a belt of thrust faulting with an NE–SW direction of the axis of maximum compression. It runs along the eastern coastline of the Adriatic Sea and terminates just north of Lefkada Island. The KTFZ is a major dextral strike slip fault zone that frequently accommodates strong earthquakes, clustered in space and time probably due to the stress transfer between the fault segments comprised in the fault system [1]. The northernmost fault segment of the KTFZ was activated in 2003 with an M_w6.2 main shock and its rich aftershock sequence. The accurate location of these aftershocks provided for the first time the indication that the major fault segments bound the western coastlines of the island, and the contemporaneously activated secondary fault segments lie onshore [2]. No earthquakes with M ≥ 6.0 have occurred during the instrumental era along the collision front in the vicinity of the activated area. Smaller magnitude earthquakes are also sparse in comparison with the remarkably active KTFZ.

Geophysical and geodetic studies have provided sufficient knowledge of the seismotectonic properties of the area. Identifying in more detail the mechanism of transition along active boundaries from the one plate to its neighbor remains a major scientific challenge. Tracing the kinematic and dynamic variability along the active boundary can provide important clues for the transition area. Microseismicity is mainly concentrated along the KTFZ and at the same time manifests the swarm-type seismogenesis as a preferential style of strain release (Figure 1). This latter property becomes the tool to decipher the localized strain pattern, since strong earthquakes occur rarely and not all in optimally oriented faults, a fact that results in appreciable gaps in seismicity catalogs both in time and space. The investigation of this activity is challenging to shed light on the seismogenic setting and the complex geodynamics of the area, a prerequisite for any seismic hazard assessment study. Its location is critical to decipher how these secondary structures play a role in accommodating strain at the borderline of the KTFZ, an area where also reverse focal mechanisms were determined for moderate magnitude earthquakes that occurred in the last few decades.

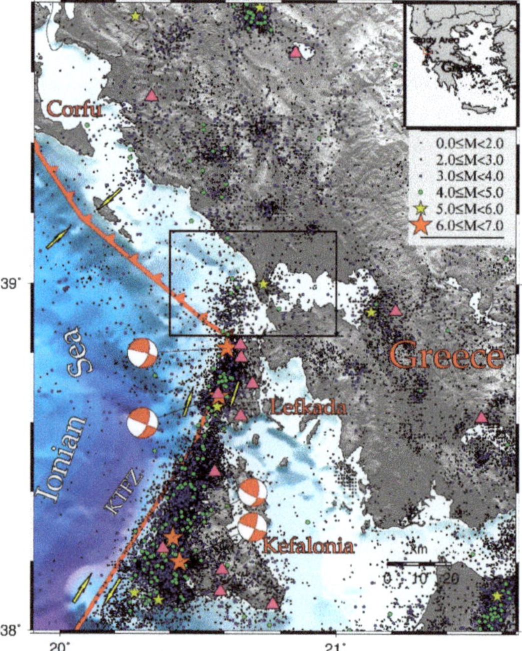

Figure 1. Active boundaries and seismicity in the Ionian Islands. The study area is delimited by a rectangle, also shown in the inset where the entirety of Greece is depicted. Earthquake epicenters are plotted according to the symbols and colors shown in the legend. The fault plane solutions of the most recent (since 2003) main shocks are shown as equal area lower hemisphere projections, where the compression quadrants are in red. The pink triangles denote the seismic stations used to record the relocated earthquakes in the current study.

The adequacy of proper data provided by an earthquake catalog compiled for the targets of this study allowed the monitoring of temporal variations of the *b* value, which from both laboratory experiments [3,4] and real data [5] seems to be linked with the stress state in a region in a way that higher differential stress corresponds to lower *b* values. We then extend our investigations to the temporal variations in *b* values aiming to detect whether they can act as an indicator for the main shock occurrence or not.

2. Seismicity Relocation

2.1. Data Acquisition and Earthquake Relocation

Since the occurrence of the 2003 Lefkada M_w6.3 main shock, continuous seismic monitoring was intensified in the area of the central Ionian Islands, namely Lefkada and Kefalonia, by firstly installing temporary broadband and short period digital seismological stations, then by upgrading this network into a permanent one [6] (http://geophysics.geo.auth.gr/ss/) and finally incorporating the stations in the Hellenic Unified Seismological Network (HUSN). For the purpose of this study, we used some additional stations located in the mainland, at distances not exceeding 90 km, in order to reduce the azimuthal gap in the events location.

In order to identify the activated structure of the northern end of the Kefalonia Transform Fault Zone (KTFZ), we compiled a catalog of 662 earthquakes, for the period February 2018–October 2019. The starting time of the catalog was decided to be a year before the main shock in order to study any pre-seismic patterns because it was observed that the seismic activity was considerable. The locations were made by manual picking of 3560 P- and 3197 S- phases, using 14 broadband and short-period stations of the HUSN. Phase pickings for the earthquakes with M ≥ 2.5 were performed manually by the analysts of the Geophysics Department of the Aristotle University of Thessaloniki (GD-AUTh). For the smaller magnitude earthquakes, phase picking was held manually by the authors of the present work, as an ongoing procedure of more deeply understanding the seismotectonic and geodynamic properties of the central Ionian islands.

Relocation was performed with the HYPOINVERSE [7] program using the 1D model (Table 1) along with the v_P/v_S ratio for this region suggested by [8], for the relocation of the 2015 Lefkada aftershock sequence. Station delays were calculated for further refining the locations, following a procedure described in [9], to account for lateral crustal variations that are not considered in a 1D model. The medians of the hypocentral errors derived by HYPOINVERSE are of the order of 2900m for the horizontal and 1500 m for the vertical dimension.

Table 1. 1D velocity model used for the seismicity relocation in the present study.

V_P (km/s)	Depth (km)
5.850	0.0
5.870	1.0
5.980	2.0
6.635	6.0
6.490	8.0
6.525	9.0
6.560	11.0
6.580	13.0
6.625	21.0
6.700	28.0
8.000	40.0

Further relocation took place by applying the double difference (DD) algorithm [10], taking into account both catalog and cross-correlation differential times and employing the conjugate gradients method (LSQR) [11] with appropriate damping. For the cross-correlation differential times, the time domain cross-correlation technique was realized [12,13]. Waveforms with a duration of 60 s, starting at the origin time of each earthquake, were prepared and band pass-filtered at 2–10 Hz with a second order Butterworth filter. Cross-correlation was performed for a 1-s window length starting at the arrival of the P and S phases, searching over a lag of 1 s. Differential times with a correlation coefficient of 0.8 or more (CC $\geq$ 0.8) were kept being used in hypoDD.

In the application of the DD algorithm, five sets of iterations with five iterations each were performed. The cross-correlation differential times were down weighted by a factor of 100 for the first 10 iterations, to obtain locations from the catalog data [8,14]. The catalog data was downweighted by the same factor for the remaining 15 iterations to allow cross-correlation differential times to better define the active structures. For the first five iterations, and considering the catalog data alone, weighting was not applied. The relocation procedure took place for interevent distances less than or equal to 5 km for the first 10 iterations of catalog data and cross-correlation times, reaching at 3 km for the cross-correlation dataset at the final set of iterations.

2.2. Location Errors Estimation

The location errors calculated by hypoDD are underestimated, usually being a few tens of meters. Thus, a bootstrap method was applied to the final residuals, creating a dataset by drawing samples from the observed residual distribution, adding them to the differential times and then repeating the relocation [15,16]. This procedure was repeated 200 times and then the 95% error ellipsoid per event was computed by the resulting catalogs.

The jackknife method [15] was also applied to evaluate the station distribution effect on the final catalog. The relocation was repeated omitting one station at a time, followed by a calculation of the standard deviations per event. The medians calculated for the axis of the ellipsoids produced by the bootstrap method and the medians of the standard deviations calculated from the jackknife method are summarized in Table 2.

The errors produced by both methods show that in the west–east horizontal direction, they are much larger than in the north–south direction. This reveals the effect of the station distribution on the location accuracy, as the closer stations are located south of the activated area in the Lefkada and Kefalonia islands, in a rather narrow N–S oriented zone covering a small azimuthal range in the E–W direction. In addition, the number of stations is small to the east and north of the study area, in western Greece, and obviously there are no stations to the west, an area covered entirely by sea.

Table 2. Median errors computed by the 95% confidence error ellipsoid for the bootstrap method and median errors of standard deviations obtained from the jackknife method. Errors are presented in the three main spatial directions.

Direction	W-E	N-S	Depth
Bootstrap Median error (m)	1447	422	1806
Jackknife Median St. Deviation Errors (m)	1055	235	330

3. Fault Plane Solutions Determination

A centroid moment tensor for the main shock of the sequence was published soon after by GCMT (https://www.globalcmt.org/CMTsearch.html), suggesting an almost pure strike slip motion with the two nodal planes steeply dipping and striking either N–S or E–W. None of the nodal planes agree with the epicentral distribution in the area surrounding the main shock. In the framework of this study, the first polarities were exploited with the use of the FPFIT program [17]. The recordings of the HUSN stations for earthquakes with magnitudes $M_L \geq 3.0$ were used for polarities identification. Since most of the stations are located south of the activated area, providing a poor azimuthal coverage, only impulsive onsets were used for the focal mechanism determination. Information on eight focal mechanisms that were calculated in this study is provided in Table 3. The available recordings in all but one solution were in the azimuthal gap range of 122°–138°, with a median equal to 128°.

One of the nodal planes determined for the main shock of M_w5.4, listed fourth in Table 3, striking at 227°, dipping to the northwest at an angle of 45°, shows a good agreement with the epicentral distribution of the relocated seismicity, as this will be discussed in the next section. For the four strongest earthquakes, the available close opposite polarities projections ascertain the robust constraint of the solution (Figure 4d). The significant dextral strike slip component (rake = 160°) agrees well with the regional stress field pattern. Most of the focal mechanisms determined for the smaller magnitude earthquakes that belong to the same cluster along with the main shock (numbered 1, 2, 3 and 5 in Table 3) are similar to that of the main shock. The fault plane solutions of the stronger earthquakes that occurred south and west of the main cluster, and impressively all of them following the stronger events occurrence of the main cluster, exhibit differentiation of the first nodal plane strike (10° and 30° for the events numbered 6 and 7 in Table 3 that belong to the second cluster, and 18° for the eighth earthquake belonging to the southernmost cluster). They also differ from the gross faulting properties of the Lefkada segment of the KTFZ [8,18].

Table 3. Information on the fault plane solutions determined in the present study. The first column gives the serial number of each solution ordered according to the occurrence date and time, shown in the second and third columns, respectively. The relocated epicentral coordinates and focal depths follow in the next three 5th to 7th columns and the magnitude in the 8th column. The strike, dip and rake of the 2 nodal planes are shown in columns 8–10 and 11–13, respectively. The azimuthal gap for each solution is given in the last column.

							Nodal Plane I			Nodal Plane II			
n	Date	Occurrence Time	Lat (°N)	Lon (°E)	Depth (km)	M_L	Strike (°)	Dip (°)	Rake (°)	Strike (°)	Dip (°)	Rake (°)	Az. Gap
1	13/01/19	21:07:21.16	38.9849	20.6569	17.24	3.6	240	76	180	330	90	14	181
2	15/01/19	01:25:05.00	38.9904	20.6587	17.34	4.5	220	45	155	328	73	48	127
3	15/01/19	02:58:46.96	38.9957	20.6517	17.90	3.7	215	40	150	329	71	54	127
4	05/02/19	02:26:09.27	39.0101	20.6870	17.86	5.2	227	45	160	331	76	47	138
5	05/02/19	09:25:28.43	38.9931	20.6636	17.35	3.3	226	67	−153	125	65	−25	131
6	06/02/19	05:30:05.52	38.9242	20.6190	13.03	3.8	10	67	−156	270	68	−25	122
7	24/02/19	21:58:30.28	38.9005	20.6110	8.34	3.8	30	55	120	165	45	54	127
8	26/02/19	10:05:59.39	38.8823	20.6103	9.18	4.0	18	62	139	130	55	35	129

4. Seismicity Temporal and Spatial Characteristics

4.1. Temporal Distribution of Earthquakes

The seismic excitation attracted our attention from the first days, given that it started on 13 January with an M3.6 earthquake, followed by numerous smaller ones, then an M4.5 shock on 15 January, followed by an M3.7 after an hour and a half (see Table 3 for occurrence dates and times), all of them felt by the inhabitants of the surrounding urban areas. The activity continued for months, exhibiting alteration of active and relatively quiet periods. In order to investigate and analytically describe the temporal behavior of this rather complex seismic activity, the cumulative number and magnitude distribution of the earthquakes as a function of time are investigated and discussed.

Figure 2a shows the cumulative number of earthquakes that occurred since February 2018. The total period is divided into four subintervals separated by three turning points where the color in the symbols is changed. The first period spans from February 2018 until 15 January 2019, where a magnitude 4.5 earthquake occurred close to the incoming main shock. The activity of the first period (green symbols in Figure 2a) presents two small swarms in February–March and May–June 2018 (temporal clustering in Figure 2b), alternating with an almost stable seismicity rate, which could be considered as the background rate (Figure 2b). The occurrence of the M4.5 earthquake on January 15 2019, along with its offsprings in one single day, is impressive (blue symbols in Figure 2a), as impressive is the relative quiescence for about 20 days, which is more obvious in Figure 2b, when on 5 February, the main shock with $M_w5.4$ occurred. One hundred aftershocks followed in two days and the activity presented an exponential-like decrease up to shortly before the occurrence of the M4.0 earthquake on 26 February. It seems that its occurrence followed a rejuvenation that could be considered as a foreshock phase, and then its imminent aftershocks created a new short seismicity intensification. Soon after, the activity declined as a function of time, as expected for an aftershock sequence.

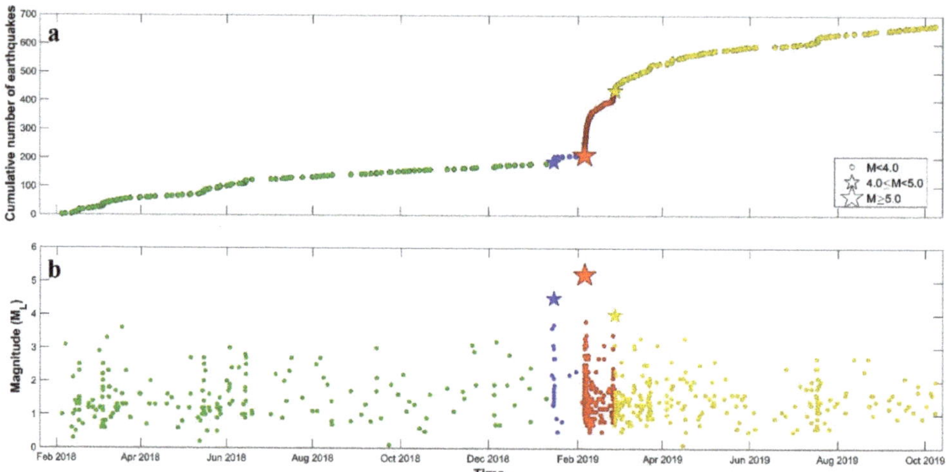

Figure 2. (**a**) Cumulative daily number of earthquakes as a function of time. A very small onset in seismicity is observed in Mid–January 2019 followed by the main onset after the main shock in early February 2019. (**b**) Earthquake magnitudes as a function of time. The small onset of seismicity in mid-January (blue symbols) corresponds to a magnitude 4.5 earthquake occurring near the main shock.

A map of the relocated seismicity is shown in Figure 3, where four colors are used to illustrate the epicenters, corresponding to the four different periods as defined in Figure 2. The seismicity is sparse and evenly distributed in the study area during 2018 (green color epicenters in Figure 3). The M4.5 earthquake triggers significant microseismicity close to its epicenter and in an area of several kilometers around (blue colored epicenters in Figure 3), much longer than its source dimensions. The $M_w5.4$, strongest earthquake in the seismic excitation, triggered aftershocks in a rather large seismic zone having an NE–SW orientation (red epicenters in Figure 3) for a short time interval during which the activity was considerably increased (red dots in Figure 2b). In one day, the activity expanded in the entire study area, manifesting a remarkable cluster comparatively far from the main shock, to the southwest of its aftershock area. The earthquake magnitudes in this cluster are all smaller than 4.0.

In less than twenty (20) days of regular aftershocks rate decay, the activity was appraised again on 26 February, when an M4.0 earthquake followed (yellow star in Figures 2 and 3). This indicates a migration and expansion of the activity to the southern part of the study area. It is of paramount importance then to explore further the sequential occurrence of the stronger members of this activity and the possible triggering through transfer of stress among the activated fault segments, which will be given in a later section of this study.

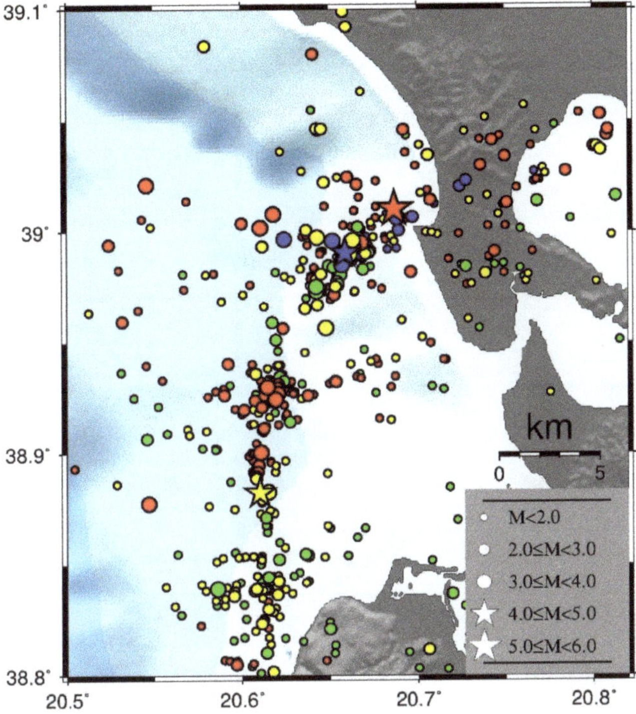

Figure 3. Spatial distribution of the relocated seismicity. The epicenters are colored according to the period of their occurrence as these periods have been defined and shown in Figure 2.

4.2. Spatial Distribution

The relocated seismicity allowed a refined and more reliable picture than the disperse epicentral distribution provided from the regional catalog as shown in Figure 1, revealing four distinctive clusters along with the disperse activity (Figure 4a). The northernmost cluster is associated with the main shock epicenter, the location of which is shown with a green star, occupying an area of almost 7km in length, slightly larger than expected for an $M_W5.4$ earthquake, according to well-accepted scaling laws [19–21]. The epicentral elongation at an NE–SW direction agrees with one of the nodal planes of the focal mechanisms of the main shock and the four aftershocks that belong to this cluster (Figure 4a and Table 3).

The activated fault segment with a strike of 220° dips to NW at an angle of 45° and is associated with dextral strike slip faulting (rake = 160°) with a considerable thrust component. This agrees with the dextral strike slip motion of the KTFZ to the south of the study area, and the continental collision to the north. The almost E–W azimuth of the axis of maximum compression also agrees with the regional stress pattern. For a more detailed investigation of the geometry of this fault segment, the relocated epicenters of the specific cluster alone are plotted in Figure 5a, along with two cross-sections, strike parallel and strike normal that are shown in Figure 5b,c, respectively. It is worth to note here that the focal depths are restricted in depths between 14 and 20 km, with the vast majority between 16 and 18 km, an important result for the identification of the seismogenic layer in the study area. The dip of the fault plane as derived from the fault plane solution is again in perfect agreement with geometry of the activated fault segment as this is revealed by the strike normal cross-section in Figure 5c.

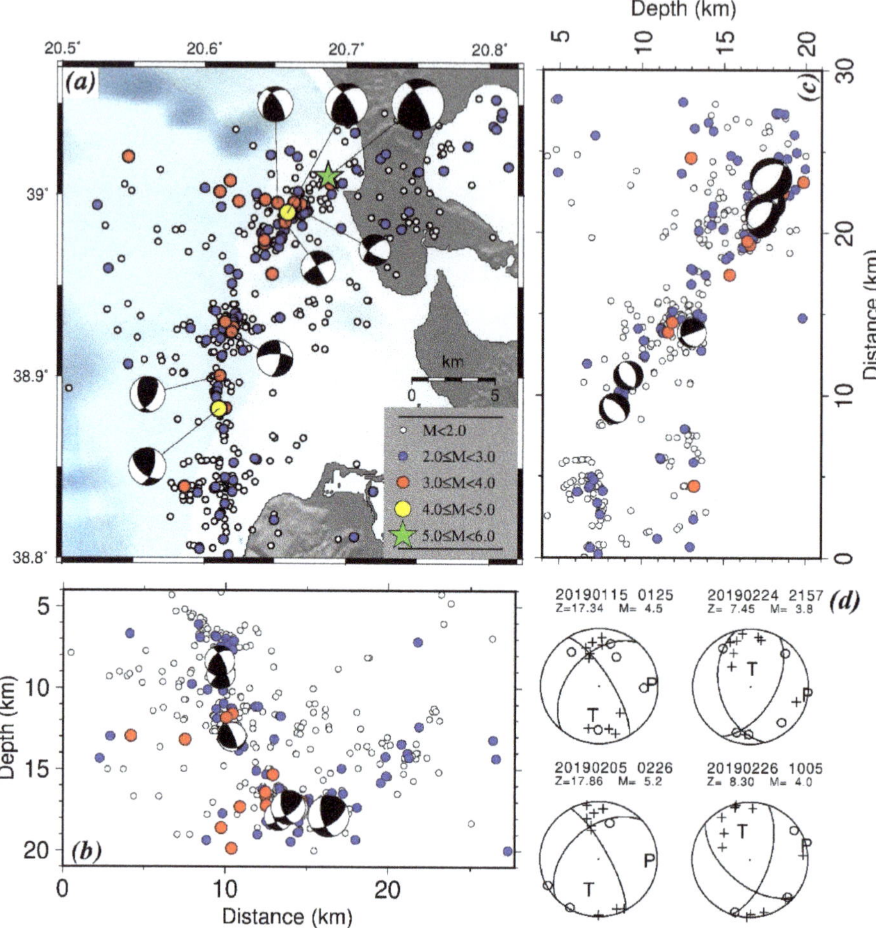

Figure 4. (**a**) Epicentral distribution of the relocated seismicity, plotted with different symbols and colors according to their magnitude as shown in the inset, along with the fault plane solutions determined in the current study depicted as equal area lower hemisphere projections, with the compression quadrants shown in black. (**b**) Cross-section across a N–S direction, containing the entire set of the relocated seismicity, with the hypocenters depicted with the same symbols as in (**a**) and the fault plane solutions as frontal projections. (**c**) Same as in (**b**) for a cross-section across a W–E direction. (**d**) Fault plane solutions along with the associated polarities ("+" for compression and "o" for dilatation) for the earthquakes numbered 2, 4, 7 and 8 in Table 3.

The next seismicity concentration is located to the south of the previous cluster and encompasses several earthquakes, the spatial distribution of which shows a barely preferable E–W elongation, in agreement with the strike of the one nodal plane from the 05/02/2019 earthquake of M3.3 (Figure 4a). The depth range is between 10 and 15 km, shallower than and exactly above the activated layer in the main cluster and with a steeper dip angle (Figure 5b,c). The faulting type, in addition to the rotation of its strike more to the north as mentioned before, exhibits a dextral strike slip motion with a slight normal component.

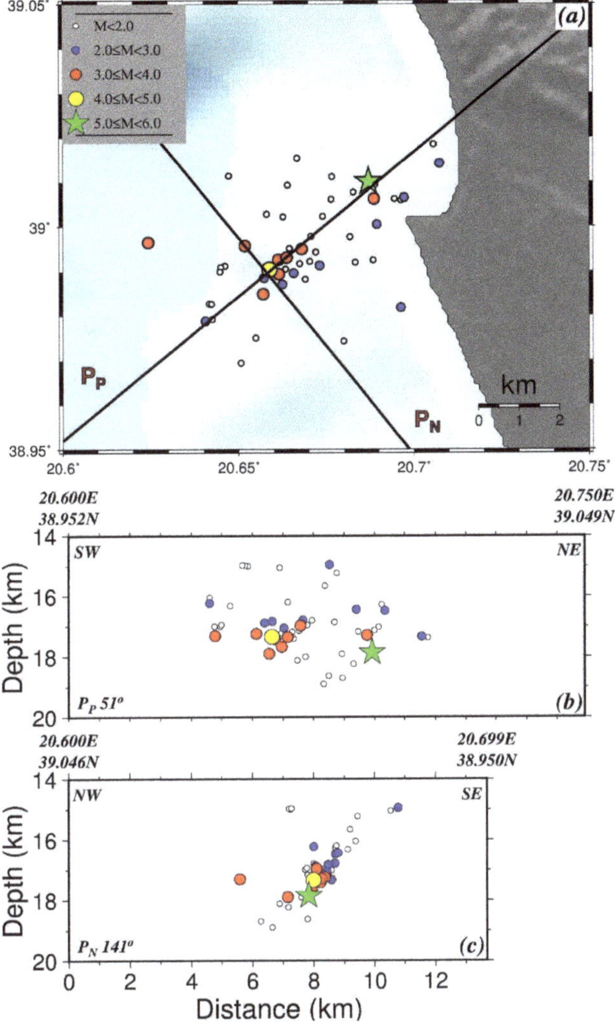

Figure 5. (a) Epicentral distribution for the main cluster, located at the northernmost point of the activity, along with two cross-sections parallel and normal to its strike, PP and PN, respectively, (b) strike parallel cross-section, (c) strike normal cross-section of the main cluster.

The epicentral distribution delineates a distinctly N–S striking structure, as one proceeds more to the south, where two sub-clusters might be recognized, forming an almost N–S striking-activated structure: the northern one, hosting the M4.0 strong event (yellow circle in Figure 4a) and forming a narrow strip of relocated epicenters, and one to its south where the activity appears sparser. The focal depths are now even more shallow, between 5 and 11 km, than the depth range of the relocated seismicity further south to Lefkada [8,22]. Almost N–S striking, dipping to the east, a dextral strike slip faulting holds for the two focal mechanisms determined for the M4.0 earthquake and one of its foreshocks, again in full agreement with the faulting properties of the Lefkada fault branch of the KTFZ [2].

5. Coulomb Stress Changes and Possible Triggering

The spatial aftershock distribution is known to be affected by the static stress changes caused by the main shock [23]. The change in Coulomb failure function (ΔCFF) is used to quantify closeness to failure [24]. A positive value of ΔCFF on a fault indicates increased likelihood that this fault will rupture in an earthquake. Values as small as 0.1 bar have been repeatedly shown to influence aftershock activity ([23,25] among others). ΔCFF depends on the changes of shear and normal stress, $\Delta\tau$ and $\Delta\sigma$, respectively, along with changes in the fluid pore pressure Δp of the rupture area, expressed by the relation

$$\Delta\text{CFF} = \Delta\tau + \mu\,(\Delta\sigma + \Delta p), \tag{1}$$

where μ is the coefficient of friction. Δp is expressed by the relation [25]

$$\Delta p = -B\,(\Delta\sigma_{kk})/3, \tag{2}$$

if undrained conditions are considered during the coseismic phase. B is the Skempton's coefficient ($0 \leq B < 1$), with experimental values between 0.5 and 0.9 shown to be appropriate for rocks [26], and $\Delta\sigma_{kk}$ is the trace of the induced stress tensor. Assuming a ductile fault zone enclosed by a homogeneous and isotropic medium, Equation (1) becomes

$$\Delta\text{CFF} = \Delta\tau + \mu'\,\Delta\sigma, \tag{3}$$

where μ' is the apparent coefficient of friction [24], which contains the effects of pore fluid as well as the material properties of the fault zone. It is estimated by the equation

$$\mu' = \mu(1 - B) \tag{4}$$

For this study, the values used to compute ΔCFF are μ=0.75 and B=0.5 as in [27] and as adopted in other studies in Greece [28,29]. These values result in an apparent coefficient of friction of $\mu'\approx$0.4, suggested by [1] for the area of the central Ionian Islands after testing different values and finding the most appropriate one for this estimation. We calculated the stress changes due to the coseismic slip of the main shock with the causative fault being approximated as a planar area with a length of 7 km and a width of 3 km, equal to the width of the seismogenic layer as explained in the previous section. A seismic moment of $M_o = 1.68 \cdot 10^{17}$ N·m was adopted from the GCMT solution and with the above source dimensions and a shear modulus of 33 GPa, a mean slip of 0.1745 m was estimated. The Poisson ratio was fixed equal to 0.25.

The Coulomb stress changes were calculated for the faulting type of the main shock (strike = 227°, dip = 45°, rake = 160°) at a depth of 16.5 km, which is approximately the middle of the seismogenic layer (Figure 5c), and are shown in Figure 6a along with the relocated aftershocks. It is evident that the vast majority of the off-fault activity is located into stress enhanced areas. Given that the faulting type of the earthquakes encompassing the other clusters differs from the one of the main shock, and that the activity was more shallow, the stress field is inverted according to the faulting type of the M4.0 shock (strike = 18°, dip = 62°, rake = 139°) at a depth of 11.5 km (Figure 6b). The earthquakes that occurred south of the main cluster are located inside areas of positive stress changes, an evidence for possible triggering, although with smaller values than in Figure 6a, possibly due to the difference in the calculation depth.

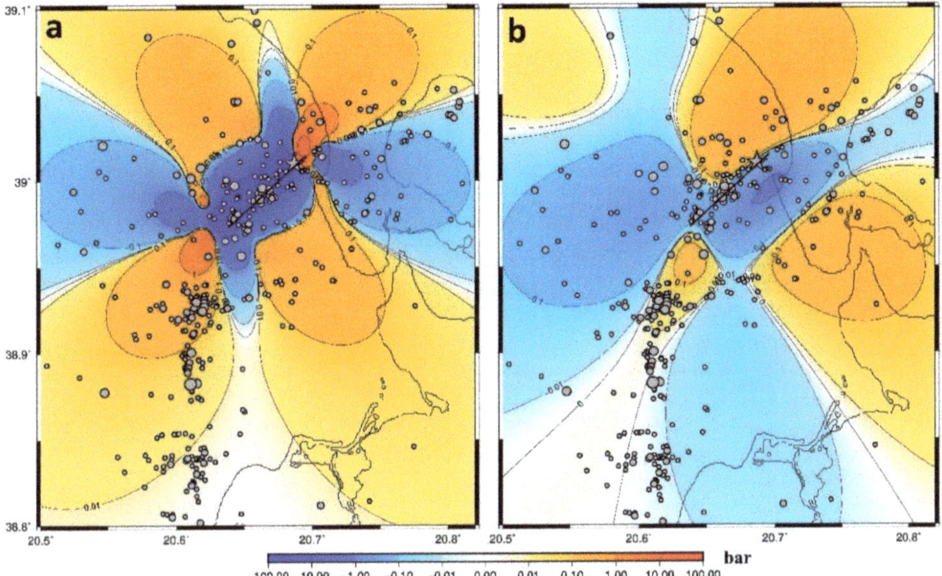

Figure 6. (**a**) Coulomb stress changes caused by the coseismic slip of the main shock, calculated at a depth of 16.5 km, according to the faulting type of the main shock. Thin black line is the fault trace at the depth of the top of the seismogenic layer, (**b**) same as in (**a**) for the faulting type of the strongest M4.0 aftershock at a depth of 11.5 km.

6. Aftershock Sequence Analysis in Relation with Temporal Variations of *b* Value

The evolution of an aftershock sequence is among the key factors for operational forecasting, given that the areas already hit by the main shock are continuously under the threat of a strong aftershock or even a higher magnitude earthquake. Monitoring the seismicity is the means to evaluate the imminent seismic hazard and several powerful tools have been developed in this respect. The occurrence of a strong aftershock, or of a larger magnitude earthquake that is afterwards considered as the main shock, along with the occurrence frequency of aftershocks depend upon the stress residuals and stress redistribution before and after the target earthquake. The direct manifestation of the stress state is the seismicity and one approach to detect fluctuations in its temporal behavior might be through the detection of the temporal variations in the *b* value. The disadvantages of this procedure are frequently connected with the lack of adequate data before the main shock that could prevent us from a robust estimation of the *b* value, and the overlapping of the coda-waves in the first hours after the main shock that can limit the early aftershocks detection capability [30]. However, the duration of our relocated catalog starting one year before the main shock and careful inspection and manual phase picking for aftershocks augmented the data sample and diminished the possible uncertainties in the *b* value estimations. The approach that we apply is described in the next steps and follows the procedure of [31].

A catalog encompassing 662 earthquakes that occurred from 4 February 2018 up to 10 October 2019 consists of our data sample for investigating the temporal behavior of the *b* value before and after the main shock that occurred on 5 February 2019, one year after the incipience of the catalog. Firstly, the completeness magnitude is detected through the maximum curvature (MAXC) method [32], which is insensitive to sample sizes and shows great stability in estimating M_C [33]. The magnitude bin with the highest frequency of events corresponds to the threshold value, which in our case is found to be $M_c = 1.3$ (Figure 7). The b value of the GR law is computed through the maximum likelihood estimation (MLE) method [34]:

$$b = \frac{1}{\ln(10)\left(\overline{M} - M_c + \Delta M/2\right)},$$

(5)

where $\overline{M}$ is the mean value of the magnitudes and ΔM is the magnitude bin, usually chosen to be equal to 0.1. This relation evidences the dependence of the b value on the choice of M_C. Figure 7b reveals a lack of earthquakes with magnitudes $M \geq 3.2$ compared with expected ones from the FMD fit. Similar breaks in FMD, also known as the nonlinearity problem, are attributed to aseismic stress release [35] as well as to a lack of faults in a particular area capable to fail in specific magnitude earthquakes [36]. In our case, the observed deficit could be related either to the seismotectonic characteristics of the study area or to the short time period of our dataset, or perhaps to both reasons. Nevertheless, the investigation of the physical mechanism responsible for the nonlinearity in the FMD remains an open issue for investigation and we believe that the data adequately support the investigation of the temporal variations in the b value for the specific seismic sequence.

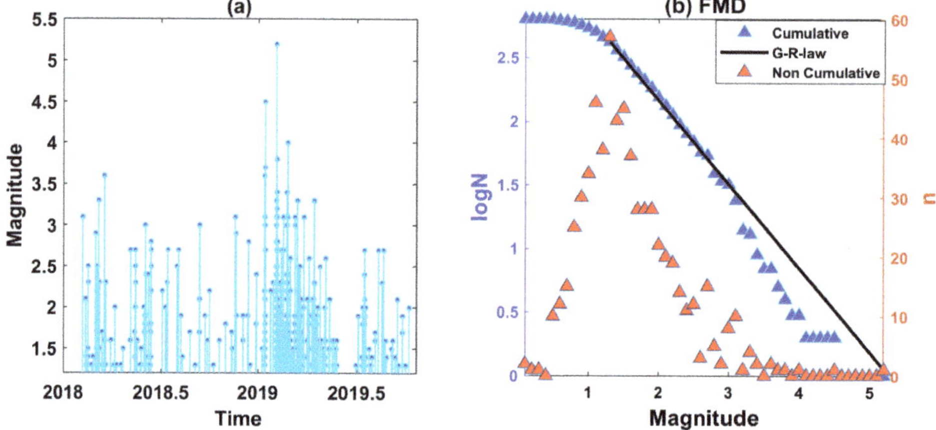

Figure 7. (a) Earthquake magnitudes, $M \geq M_c$, as a function of time, (b) incremental (red triangles) and logarithm of the cumulative frequency (blue triangles) as a function of magnitude. The black line is the GR law fit with $M_c = 1.3$ according to maximum curvature (MAXC).

For the assessment of the temporal variations in the b value before and after the main shock, firstly, we divided the catalog into two sub-catalogs, before and after the main shock with the N_{pre} and N_{after} sample sizes, respectively. Then, for each dataset, we recomputed the magnitude of completeness with the same method, since it is more appropriate than the goodness of fit method for small sample sizes ($N^* \leq 200$), which is the case here. A correction factor of +0.2 is considered in order to avoid biased results in the b value estimation, since the M_c tends to be underestimated [33,37]. Table 4 summarizes the details of the two datasets. In both cases, the magnitude bin with the highest frequency corresponds to $M = 1.3$, so the completeness magnitude after the addition of the correction factor +0.2 is $M_c = 1.5$, leading to $N_{pre} = 94$ events before the main shock and $N_{aft} = 226$ aftershocks (see Figure S1 for the GR law distribution of both datasets). Then, a sample window of size n_{pre} and n_{aft} is determined for the datasets before and after the main shock, respectively, for which we iteratively estimated the b values moving at each iteration one event forward. A minimum number of events for the sample size $n_{thr} = 50$ is defined, which is the threshold for an approximately 15% error in the estimation of the b value [35]. In this way, we achieved statistically reliable results.

Table 4. Information on the two datasets used in the study.

Datasets	Time Interval	N	M_c	N_c	a	b
Pre-main shock	4/2/2018–5/2/2019	122	1.5	94	2.84	0.58
After-main shock	5/2/2019–8/10/2019	298	1.5	226	3.48	0.75

We tested different sample sizes in order to examine the dependency of the results on the choice of the free parameters, n_{pre} and n_{aft} (Figure 8). The time-varying evolution of the b value does not significantly depend on the choice of these parameters in the size ranges (50 , 85). We set $n_{pre} = 50$ and $n_{aft} = 70$, with 45 and 157 number of samples, respectively. However, we stress that with any other combination, the same conclusions would have been reached.

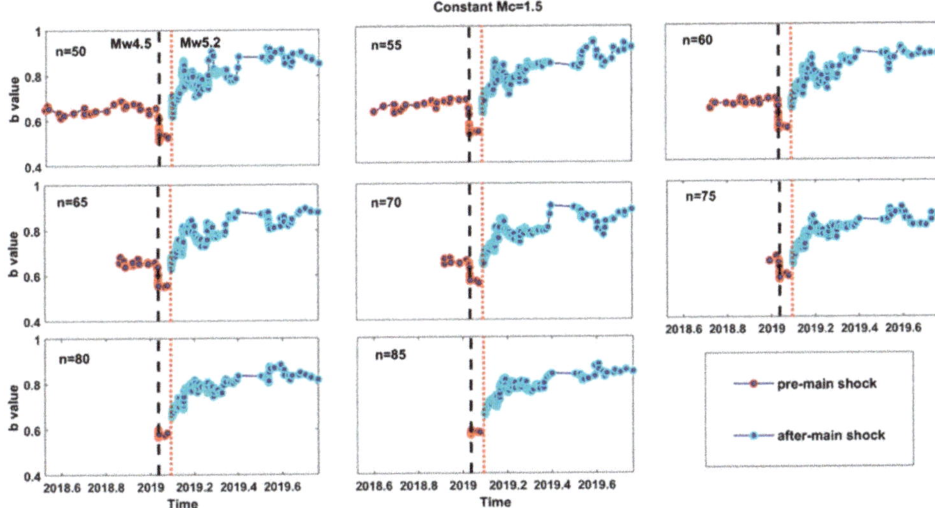

Figure 8. Time-series of the b value for different sample sizes. The number of shocks (n) of each sample is written at the left top corner of each box and it is the same for both the pre-main shock and after-main shock datasets. Red dotted line corresponds to the main shock occurrence time, black dashed line to the foreshock $M4.5$ occurrence time and the red and blue dots are plotted at the occurrence time of the last event in each sample, respectively.

In our case, the magnitude completeness is assumed constant, however, it is generally suggested to reevaluate it at each step since it can change in time, due to seismic network modifications or detection limitations. We keep the same initially estimated completeness threshold for two reasons. Firstly, the dataset covers a short period with relatively few events since it concerns the aftershock sequence of a moderate size main shock, which must be taken into account with caution for a robust estimation of the b value while evaluating iteratively the completeness threshold (many samples fall under the minimum value n_{thr}). Secondly, we verified that the geometry of the seismic network during the study period was stable, meanwhile, we are highly confident about the monitoring analysis quality since it was under our surveillance for the entire time interval. Nevertheless, we tested the temporal fluctuations of the b values both with constant and moving magnitude completeness thresholds and we did not observe significant variations through time (Figures S2 and S3). In all cases, the main observation of decreasing b values after the $M4.5$ foreshock (dashed black line in Figure 8) compared to the increasing ones just after the main shock held true, therefore, in our analysis, we continued with a constant threshold that let us include more samples.

Finally, we computed the median of the *b* values for the period of almost one year (from February 2018 until the last event before the foreshock of *M*4.5) as a reference value for the comparison with the period between the foreshock and the main shock that occurred 21 days later, as well as with the entire aftershock period. In Figure 9, the temporal variation in the *b* values with $b_{ref} = 0.64$ is illustrated. We can see that after the foreshock of *M*4.5 on 15 January 2019, the *b* values decrease rapidly compared with the reference value by approximately 10%. In contrast, a general increase in the *b* values after the main shock occurrence is observed, that exceeds 10% a day after, continues to increase and remains at a level reaching 20% or more up to the end of the study period.

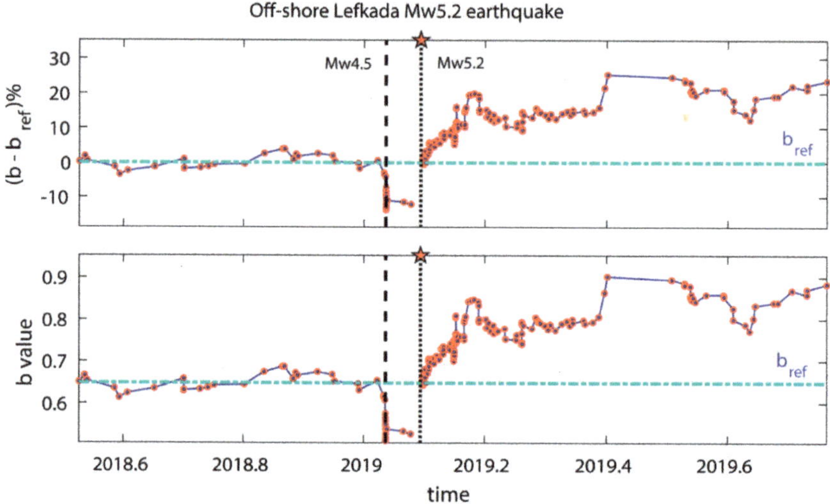

Figure 9. Top: the percent difference in the *b* values (blue circles) compared with the reference value (cyan dashed-dot line). The vertical dotted line with the red star corresponds to the main shock occurrence time and the black dashed vertical line to the occurrence time of the foreshock. Bottom: same as in top but for the estimated *b* values.

7. Discussion and Conclusions

The seismic activity that manifested in 2019 in the offshore area north of Lefkada Island motivated our investigation for the identification of the geometry and kinematics of the activated fault segments, given that they are located at the junction between the two major active boundaries, namely the dextral strike slip Kefalonia Transform Fault Zone and the collision between the Adriatic microplate and the Greek mainland. It was expected to unveil the mode of kinematics and fill in the gap for the transition from transform to collision, adding value to the exhaustive relocation and manual picking of arrival phases for as many earthquakes as possible of the recorded ones, and the first polarities of the stronger shocks that are associated with the activated fault segments.

The exact location and geometry of the offshore part of the KTFZ and its termination against the thrust front was not identified with seismological data up to now. This lack of knowledge about its northern extent is mostly due to the lack of moderate earthquakes at this location during recent years where the seismological network detectability became efficient to provide the data for this analysis. The geometry of the offshore faults identified in this study exhibiting mostly strike slip motion do not match the orientation of the Lefkada fault branch. The difference in strike is ~20°, with the incorporation of a small thrust component in agreement with the adjacency of the compressional stress field regime. The three observed clusters that are traced from the relocated seismicity in this study exhibit spatial properties differentiation and reveal the 3D local structure, given that the focal depths were relocated with high accuracy. The northern fault segment, although traced in already suggested kinematic models [38,39], is for the first time to be identified and investigated with seismological recordings, and in particular, with such fine details. [38] referred to a right lateral strike-slip system that resulted from the collision of western Greece with the Apulian platform, and runs through western Greece to the KTFZ. The position of the identified activated fault segments, their strike and sense of slip agree with the active fault mapping and suggested regional tectonics of [40]. A clockwise rotation of the Ionian Islands and Akarnania block accommodated by major marginal strike-slip zones that appeared segmented along their strike was suggested by [41].

Available fault plane solutions of moderate and strong earthquakes (M ≥ 5.0) in the Lefkada segment of the KTFZ and the collision front extend as far as in the past such as 1973, from various sources ([42] for the earthquake of 1973/11/04, [43] for the earthquake of 25/02/1994 and the GCMT catalogue). Nevertheless, there are not any available focal mechanism solutions for earthquakes of such magnitude in the study area. Fault plane solutions for smaller magnitude earthquakes in the offshore area north of and close to Lefkada Island can be found in several studies ([8,44,45] among others). Those solutions are either covering the offshore area that is the southern edge of our study area, thus consistently following the NNE–SSW strike of the main fault zone, or they are not accompanied by a highly accurate focal distribution as in our study, because even in the most recent cases of the 2003 and 2015 earthquakes, the seismic activity in the area of the main fault under study was not intense.

Figure 10 shows the identified active structures and a kinematic sketch of the study area. The focal mechanisms determined in the present study (compression quadrants shown in red, and numbering according to Table 3) along with the focal mechanisms of earthquakes with M ≥ 5.0 were already published (green compression quadrants). Their origin time (YYYY/MM/DD) is shown on the top of the beach balls. The thick gray lines denote the major active boundaries, namely the KTFZ, where the dextral strike slip motion is shown by the antiparallel yellow arrows, and the collision front by the sawtooth line. The prolongation of the KTFZ in the study area, as identified and documented in the present study, is colored in red. The inset map shows the axes of maximum stress, P-axes, as they have been taken from the fault plane solutions shown in the main part of the figure. The thick brown curved arrows follow the P-axis rotation in agreement with the strike slip motion along the KTFZ and the counterclockwise rotation of the Adria microplate.

The local fault system sheds more light on the transition from the dextral strike slip motion along the KTFZ to the thrust faulting along the collision front. The thrust component present in the fault plane solutions determined for the stronger earthquakes of the 2019 swarm put in evidence the influence of the compressive stress field to the strike slip faults that are either striking along the KTFZ (the southern ones) or "kinking" (the northern one) for compensating the kinematic direction prevalent at this location. The slight but gradually changing rotation in the azimuth of the maximum stress axes, from south (green arrows in Figure 10) to the northernmost edge of the 2019 activated structures (red arrows), supports the hypothesis of a transpression movement in a transition zone between the transform and collision zones.

The examination of the *b* value changes shows lower values before and higher after the main shock occurrence. A noticeable decrease in the *b* value (from 0.6 to 0.5) was observed immediately after the occurrence of the M4.5 foreshock on January 15 2019 and remained low until the main shock occurrence on 5 February 2019. Then, it sharply increased to the values before the M4.5 foreshock and gradually became higher, approaching the value of 0.9.

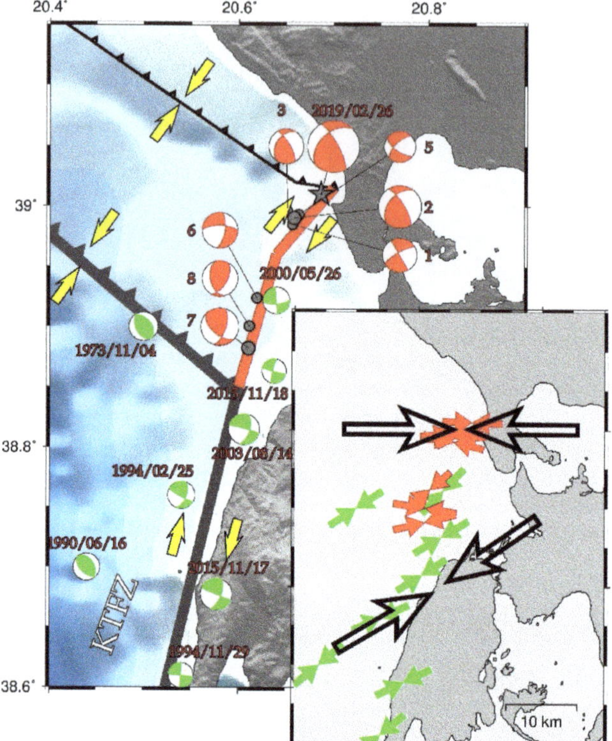

Figure 10. Fault plane solutions of 16 earthquakes projected as equal area lower hemisphere projections, where compression quadrants are in green for earthquakes of M > 5.0, that occurred before the earthquake sequence in study and in red for earthquakes that were determined and belong to the swarm investigated in this study. Thick grey lines denote the active boundaries of the Kefalonia Transform Fault Zone (KTFZ) and the Apulian collision front. Red lines depict the identified fault segment offshore Lefkada. Thin black line represents the Ionian thrust. Yellow arrows indicate relative plate motion. The inset shows the P-axis inferred from the focal mechanisms. Black arrows illustrate the characteristic direction of the axis, NE–SW for the earthquakes along the KTFZ and N–E for those.

Supplementary Materials: The following are available online at http://www.mdpi.com/2076-3417/10/13/4491/s1, Figure S1: Frequency-magnitude distribution (FMD of pre-main shock and post-main shock datasets, Figure S2: Time-series of the *b* value of different sample sizes for the pre-main shock dataset, Figure S3: Same as in Figure S2 for the post-main shock dataset.

Author Contributions: Conceptualization, A.K., E.P. and V.K.; methodology, A.K. and P.B.; software, A.K. and P.B.; validation, A.K. and P.B.; formal analysis, A.K. and P.B.; investigation, A.K. and P.B., data curation, A.K.; writing—original draft preparation, A.K. and P.B.; writing—review and editing, E.P. and V.K.; visualization, A.K. and P.B.; supervision V.K. and E.P.; project administration, V.K. and E.P.; funding acquisition, V.K.; All authors have read and agreed to the published version of the manuscript.

Funding: This research is co-financed by Greece and the European Union (European Social Fund—ESF) through the Operational Programme «Human Resources Development, Education and Lifelong Learning 2014–2020» in the context of the project "Kinematic properties, active deformation and stochastic modelling of seismogenesis at the Kefalonia–Lefkada transform zone" (MIS–5047845).

Acknowledgments: The software Generic Mapping Tools was used to plot the map of the study area [46].

Conflicts of Interest: The authors declare no conflict of interest. The funders had no role in the design of the study; in the collection, analyses, or interpretation of data; in the writing of the manuscript, or in the decision to publish the results.

References

1. Papadimitriou, E.E. Mode of strong earthquake recurrence in the central Ionian Islands (Greece): Possible triggering due to Coulomb stress changes generated by the occurrence of previous strong shocks. *Bull. Seismol. Soc. Am.* **2002**, *92*, 3293–3308. [CrossRef]
2. Karakostas, V.G.; Papadimitriou, E.E.; Papazachos, C.B. Properties of the 2003 Lefkada, Ionian Islands, Greece, earthquake seismic sequence and seismicity triggering. *Bull. Seismol. Soc. Am.* **2004**, *94*, 1976–1981. [CrossRef]
3. Goebel, T.H.; Schorlemmer, D.; Becker, T.W.; Dresen, G.; Sammis, C.G. Acoustic emissions document stress changes over many seismic cycles in stick-slip experiments. *Geophys. Res. Lett.* **2013**, *40*, 2049–2054. [CrossRef]
4. Scholz, C.H. The frequency-magnitude relation of microfracturing in rock and its relation to earthquakes. *Bull. Seismol. Soc. Am.* **1968**, *58*, 399–415.
5. Schorlemmer, D.; Wiemer, S.; Wyss, M. Variations in earthquake-size distribution across different stress regimes. *Nature* **2005**, *437*, 539–542. [CrossRef] [PubMed]
6. Permanent Regional Seismological Network operated by the Aristotle University of Thessaloniki. Available online: http://geophysics.geo.auth.gr/the_seisnet/WEBSITE_2005/station_index_en.html (accessed on 17 May 2020). [CrossRef]
7. Klein, F.W. *User's Guide to HYPOINVERSE-2000, A Fortran Program to Solve Earthquake Locations and Magnitudes*; U.S. Geological Survey Open File Report, 02-171 Version 1.0; U.S. Geological Survey: Reston, VA, USA, 2000.
8. Papadimitriou, E.; Karakostas, V.; Mesimeri, M.; Ghouliaras, C.; Kourouklas, C. The M_w6.5 17 November 2015 Lefkada (Greece) Earthquake: Structural Interpretation by Means of the Aftershock Analysis. *Pure Appl. Geophys.* **2017**, *174*, 3869–3888. [CrossRef]
9. Karakostas, V.G.; Papadimitriou, E.E.; Gospodinov, D. Modeling the 2013 North Aegean (Greece) seismic sequence: Geometrical and frictional constraints, and aftershock probabilities. *Geophys. J. Int.* **2014**, *197*, 525–541. [CrossRef]
10. Waldhauser, F. *HypoDD—A Program to Compute Double–Difference Hypocenter Locations*; US Geological Survey Open File Report; U.S. Geological Survey: Reston, VA, USA, 2001; pp. 1–113.
11. Paige, C.C.; Saunders, M.A. LSQR: An Algorithm for Sparse Linear Equations and Sparse Least Squares. *ACM Trans. Math. Softw.* **1982**, *8*, 43–71. [CrossRef]
12. Schaff, D.P.; Bokelmann, G.H.R.; Ellsworth, W.L.; Zanzerkia, E.; Waldhauser, F.; Beroza, G.C. Optimizing correlation techniques for improved earthquake location. *Bull. Seismol. Soc. Am.* **2004**, *94*, 705–721. [CrossRef]
13. Schaff, D.P.; Waldhauser, F. Waveform cross–correlation–based differential travel–time measurements at the northern California seismic network. *Bull. Seism. Soc. Am.* **2005**, *95*, 2446–2461. [CrossRef]
14. Mesimeri, M.; Karakostas, V.; Papadimitriou, E.; Tsaklidis, G.; Jacobs, K. Relocation of recent seismicity and seismotectonic properties in the Gulf of Corinth (Greece). *Geophys. J. Intern.* **2018**, *212*, 1123–1142. [CrossRef]
15. Efron, B. The Jackknife, the Bootstrap and Other Resampling Plans. *Soc. Idustr. Appl. Math. Phila.* **1982**, 5–11. [CrossRef]
16. Waldhauser, F.; Ellsworth, W.L. A double-difference earthquake location algorithm: Method and application to the Northern Hayward Fault, California. *Bull. Seism. Soc. Am.* **2000**, *90*, 1353–1368. [CrossRef]
17. Reasenberg, P.; Oppenheimer, D.H. *FPFIT, FPPLOT and FPPAGE; Fortran Computer Programs for Calculating and Displaying Earthquake Fault-Plane Solutions*; Open-File Rep. 85-739; U.S. Geological Survey: Reston, VA, USA, 1985.
18. Karakostas, V.G.; Papadimitriou, E.E. Fault complexity associated with the 14 August 2003 Mw6.2 Lefkada, Greece, aftershock sequence. *Acta Geophys.* **2010**, *58*, 838–854. [CrossRef]
19. Wells, D.L.; Coppersmith, K.J. New Empirical Relationships among Magnitude, Rupture Length, Rupture width, Rupture Area, and Surface Displacement. *Bull. Seismol. Soc. Am.* **1994**, *84*, 974–1002.
20. Papazachos, B.C.; Scordilis, E.M.; Panagiotopoulos, D.G.; Papazachos, C.B.; Karakaisis, G.F. Global relations between seismic fault parameters and moment magnitude of earthquakes. In Proceedings of the 10th International Congress of the Geological Society, Thessaloniki, Greece, 14–17 April 2004; pp. 539–540. [CrossRef]

21. Thingbaijam, K.K.S.; Mai, P.M.; Goda, K. New Empirical Earthquake Source-Scaling Laws. *Bull. Seismol. Soc. Am.* **2017**, *107*, 2225–2246. [CrossRef]

22. Ganas, A.; Elias, P.; Bozionelos, G.; Papathanassiou, G.; Avallone, A.; Papastergios, A.; Valkaniotis, S.; Parcharidis, I.; Briole, P. Coseismic deformation, field observations and seismic fault of the 17 November 2015 M = 6.5, Lefkada Island, Greece earthquake. *Tectonophysics* **2016**, *687*, 210–222. [CrossRef]

23. King, G.C.P.; Stein, R.S.; Jian, L. Static stress changes and the triggering of earthquakes. *Bull. Seismol. Soc. Am.* **1994**, *84*, 935–953. [CrossRef]

24. Harris, R.A. Introduction to Special Section: Stress Triggers, Stress Shadows, and Implications for Seismic Hazard. *J. Geophys. Res. Solid Earth* **1998**, *103*, 24347–24358. [CrossRef]

25. Reasenberg, P.A.; Simpson, R.W. Response of regional seismicity to the static stress change produced by the Loma Prieta earthquake. *Science* **1992**, *255*, 1687–1690. [CrossRef]

26. Rice, J.R.; Cleary, M.P. Some basic stress diffusion solutions for fluid–saturated elastic porous media with compressible constituents. *Rev. Geophys. Space Phys.* **1976**, *14*, 227–241. [CrossRef]

27. Robinson, R.; McGinty, P.J. The enigma of the Arthur's Pass, New Zealand, earthquake: 2. The aftershock distribution and its relation to regional and induced stress fields. *J. Geophys. Res.* **2000**, *105*, 16139–16150. [CrossRef]

28. Karakostas, V.; Papadimitriou, E.; Mesimeri, M.; Gkarlaouni, C.; Paradisopoulou, P. The 2014 Kefalonia Doublet (M_w6.1 and M_w6.0), central Ionian Islands, Greece: Seismotectonic implications along the Kefalonia transform fault zone. *Acta Geophys.* **2015**, *63*, 1–16. [CrossRef]

29. Karakostas, V.; Mirek, K.; Mesimeri, M.; Papadimitriou, E.; Mirek, J. The Aftershock Sequence of the 2008 Achaia, Greece, Earthquake: Joint Analysis of Seismicity Relocation and Persistent Scatterers Interferometry. *Pure Appl. Geophys.* **2017**, *174*, 151–176. [CrossRef]

30. Lippiello, E.; Cirillo, A.; Godano, C.; Papadimitriou, E.; Karakostas, V. Post Seismic Catalog Incompleteness and Aftershock Forecasting. *Geosciences* **2019**, *9*, 355. [CrossRef]

31. Gulia, L.; Wiemer, S. Real-time discrimination of earthquake foreshocks and aftershocks. *Nature* **2019**, *574*, 193–199. [CrossRef] [PubMed]

32. Wiemer, S.; Wyss, M. Minimum magnitude of completeness in earthquake catalogs: Examples from Alaska, the Western United States, and Japan. *Bull. Seism. Soc. Am.* **2002**, *90*, 859–869. [CrossRef]

33. Zhou, Y.; Zhou, S.; Zhuang, J. A test on methods for MC estimation based on earthquake catalog. *Earth Planet. Phys.* **2018**, *2*, 150–162. [CrossRef]

34. Aki, K. Maximum likelihood estimate of b in the formula log N=a-bM and its confidence limits. *Bull. Earthq. Res. Inst.* **1965**, *43*, 237–239.

35. Tormann, T.; Wiemer, S.; Mignan, A. Systematic survey of high-resolution b value imaging along Californian faults: Inference on asperities. *J. Geophys. Res. Sol. Earth* **2014**, *119*, 2029–2054. [CrossRef]

36. Console, R.; Carluccio, R.; Papadimitriou, E.; Karakostas, V. Synthetic earthquake catalogs simulating seismic activity in the Corinth Gulf, Greece, fault system. *J. Geophys. Res. Sol. Earth* **2015**, *120*, 326–343. [CrossRef]

37. Woessner, J.; Wiemer, S. Assessing the quality of earthquake catalogues: Estimating the magnitude of completeness and its uncertainty. *Bull. Seismol. Soc. Am.* **2005**, *95*, 684–698. [CrossRef]

38. van Hinsbergen, D.J.J.; van der Meer, D.G.; Zachariasse, W.J.; Meulenkamp, J.E. Deformation of western Greece during Neogene clockwise rotation and collision with Apulia. *Int. J. Earth Sci. (Geol. Rundsch.)* **2006**, *95*, 463–490. [CrossRef]

39. Vassilakis, E.; Royden, L.; Papanikolaou, D. Kinematic links between subduction along the Hellenic trench and extension in the Gulf of Corinth, Greece: A multidisciplinary analysis. *Earth Planet. Sci. Lett.* **2011**, *303*, 108–120. [CrossRef]

40. Perouse, E.; Sebrier, M.; Braucher, R.; Chamot Rooke, N.; Bourles, D.; Briole, P.; Sorel, D.; Dimitrov, D.; Arsenikos, S. Transition from collision to subduction in western Greece: The katouna—Stamna active fault system and regional kinematics. *Int. J. Earth Sci. (Geol. Rundsch.)* **2017**, *106*, 967–989. [CrossRef]

41. Haddad, A.; Ganas, A.; Kassaras, I.; Lupi, M. Seismicity and geodynamics of western Peloponnese and central Ionian Islands: Insights from a local seismic deployment. *Tectonophysics* **2020**. [CrossRef]

42. Baker, C.; Hatzfeld, D.; Lyon-Caen, H.; Papadimitriou, E.; Rigo, A. Earthquake mechanisms of the Adriatic Sea and Western Greece: Implications for the oceanic subduction-continental collision transition. *Geophys. J. Int.* **1997**, *131*, 559–594. [CrossRef]

43. Louvari, E.; Kiratzi, A.A.; Papazachos, B.C. The Cephalonia Transform Fault and its extension to western Lefkada Island (Greece). *Tectonophysics* **1999**, *308*, 223–236. [CrossRef]
44. Hatzfeld, D.; Kassaras, I.; Panagiotopoulos, D.; Amorese, D.; Makropoulos, K.; Karakaisis, G.; Coutant, O. Microseimicity and strain pattern in northwestern Greece. *Tectonics* **1995**, *14*, 773–785. [CrossRef]
45. Benetatos, C.; Kiratzi, A.; Roumelioti, Z.; Stavrakakis, G.; Drakatos, G.; Latoussakis, I. The 14 August 2003 Lefkada Island (Greece) earthquake: Focal mechanisms of the mainshock and of the aftershock sequence. *J. Seismol.* **2005**, *9*, 171–190. [CrossRef]
46. Wessel, P.; Smith, W.H.F.; Scharroo, R.; Luis, J.; Wobbe, F. Generic Mapping Tools: Improved Version Released. *EOS Trans. Am.* **2013**. [CrossRef]

 *applied*
sciences

Review

The BrIdge voLcanic LIdar—BILLI: A Review of Data Collection and Processing Techniques in the Italian Most Hazardous Volcanic Areas

Stefano Parracino [1], Simone Santoro [2], Luca Fiorani [2,*], Marcello Nuvoli [2], Giovanni Maio [3] and Alessandro Aiuppa [4,5]

[1] Department of Industrial Engineering, University of Rome "Tor Vergata", 00173 Rome, Italy; stefano.parracino@uniroma2.it
[2] Nuclear Fusion and Safety Technologies Department, ENEA (Italian National Agency for New Technologies, Energy and Sustainable Economic Development), 00044 Frascati, Italy; simone.santoro@enea.it (S.S.); marcello.nuvoli@enea.it (M.N.)
[3] BU Space & Big Science, Vitrociset a Leonardo Company, 00156 Rome, Italy; g.maio@vitrociset.it
[4] Dipartimento di Scienze della Terra e del Mare, Università di Palermo, 90123 Palermo, Italy; alessandro.aiuppa@unipa.it
[5] Istituto Nazionale di Geofisica e Vulcanologia, 90146 Palermo, Italy
* Correspondence: luca.fiorani@enea.it

Received: 24 June 2020; Accepted: 10 September 2020; Published: 14 September 2020

Featured Application: Authors are encouraged to provide a concise description of the specific application or a potential application of the work. This section is not mandatory.

Abstract: Volcanologists have demonstrated that carbon dioxide (CO_2) fluxes are precursors of volcanic eruptions. Controlling volcanic gases and, in particular, the CO_2 flux, is technically challenging, but we can retrieve useful information from magmatic/geological process studies for the mitigation of volcanic hazards including air traffic security. Existing techniques used to probe volcanic gas fluxes have severe limitations such as the requirement of near-vent in situ measurements, which is unsafe for operators and deleterious for equipment. In order to overcome these limitations, a novel range-resolved DIAL-Lidar (Differential Absorption Light Detection and Ranging) has been developed as part of the ERC (European Research Council) Project "BRIDGE", for sensitive, remote, and safe real-time CO_2 observations. Here, we report on data collection, processing techniques, and the most significant findings of the experimental campaigns carried out at the most hazardous volcanic areas in Italy: Pozzuoli Solfatara (Phlegraen Fields), Stromboli, and Mt. Etna. The BrIdge voLcanic LIdar—BILLI has successfully obtained accurate measurements of in-plume CO_2 concentration and flux. In addition, wind velocity has also been retrieved. It has been shown that the measurements of CO_2 concentration performed by BILLI are comparable to those carried out by volcanologists with other standard techniques, heralding a new era in the observation of long-term volcanic gases.

Keywords: volcanic eruptions; volcanic plumes; CO_2 flux; DIAL-Lidar; data processing techniques

---

1. Introduction

Volcanoes represent a significant threat and, at the same time, are a huge attraction since antiquity for populations living nearby due to their fertile lands. At present, large populations live near active or quiescent volcanoes and thus are at risk of their possible eruptions. It is well known that volcanic eruptions can determine increasing air pollution levels; furthermore, they can influence climate [1] and, in some circumstances, produce lethal eruptions that destroy the surrounding environment and cause serious losses to national/international economies. Cautious valuation of volcano comportment and activity state is required to mitigate these effects, which can be carried out by instruments dedicated to volcanic control [2].

Volcanologists can nowadays infer in real time, by modern technological and modeling developments, the signals coming from seismic and ground deformations that usually happen before eruptions. It is possible to know of an imminent eruption by detecting the stress changes and magma accumulation/flow, thanks to the improvements in broadband seismometers, models, and tools for processing seismic signals [3], and the constant implementation of satellite-based (GPS, Global Positioning System [4] and InSAR, Interferometric Synthetic Aperture Radar [5]) geodetic observations [6].

Volcanic plumes, fumaroles, and degassing grounds release incessantly magmatic volatiles (e.g., H_2O, SO_2, and CO_2) [7], representing the external appearance of deep magma degassing [8,9]. These volatiles are a source of priceless data for forecasting the likelihood of a volcanic eruption. Therefore, accurate knowledge of volcanic gas composition and flux can provide alerts on possible eruptions [10], and provide insights into the geophysical processes occurring in the inner parts of volcanoes.

As far as abundancy is concerned, CO_2 is the second gas in volcanic fluids. Unfortunately, in contrast to SO_2, its measurements are much more complicated. In fact, efforts to achieve standoff detection of the volcanic CO_2 flux have been discouraged by technical challenges. This is mainly due to the large atmospheric background, about 400 parts per million (ppm), which makes it difficult to resolve the volcanic CO_2 signal. For these reasons, there are few volcanic CO_2 flux data in geophysical studies (see [11–14] for fresh research).

In fact, SO_2 can be found at the parts per billion concentration in the unperturbed atmosphere, therefore its volcanic flux can be detected via routinely collected data from the ground (e.g., DOAS, Differential Optical Absorption Spectroscopy [15]) and in space using UV (Ultraviolet) spectroscopy [16–18]. Instead, standoff detection of volcanic CO_2 has only been accomplished during eruptions of mafic volcanoes, where ground-based Fourier transform infra-red (FTIR) spectrometers can take advantage of magma/hot rocks as efficient light sources [19]. Conversely, monitoring of the far more usual "passive" CO_2 releases from quiescent volcanoes has implied access to risky summit craters for direct measurement of fumaroles [20] or in situ sampling of plumes by implementing Multi-GAS (Multi-component Gas Analyzer) systems [21] or Active-FTIR [22].

The employment of modern systems and networks for volcanic gas observations [8,20,23] has made it possible to make significant progress in direct sampling techniques for volcano monitoring. This has resulted in an improvement in time resolution of traditional ground-based volcanic gas observations [24]. Moreover, taking advantage of fast advancements in optoelectronics, especially in the last two decades including new coherent sources, detection systems, and spectrometric devices, we have at present many optical remote techniques, both passive and active, ground-based, airborne, and even space-borne, that are applicable to issues involving Earth CO_2 degassing [12,14]. The improvements of volcanic gas measurements in terms of numbers, quality, and time duration are providing evidence that CO_2 is the gas most directly linked to "pre-eruptive" degassing processes [21]. As confirmation of this fact, precursory increases in CO_2 plume flux have been recently detected at several volcanoes [25] such as at Mt. Etna, the most active and dangerous volcano in Europe [26].

Taking into account the potentialities given by data on the volcanic CO_2 flux for predicting eruptions, the ability to remotely monitor this flux with a high time resolution is an important step ahead in current volcanology.

Considering the stringent requirements of volcanological research and the great advantages of active optical remote sensing (e.g., measurement not dependent of sunlight or other radiation sources, safe monitoring distance, probing of sites difficult to access, large areal coverage, higher accuracy than passive techniques, etc. [14]), ground-based lidar systems have represented, since their creation, a valuable option for the remote detection of volcanic gases and aerosols. Such systems share many advantages with seismic and deformation monitoring with respect to direct sampling; in a nutshell, they can perform measurements from a safe area, while also allowing (semi)-continuous operation during eruptions. Moreover, gas amounts can be retrieved non-invasively in near real-time, obviating the need for successive laboratory analysis, thus avoiding possible sample contamination, and—unlike in situ sampling—standoff sensing techniques measure integrated or range-resolved gas concentrations through cross-sections of the plume, thus providing a more informative characterization of composition and flux of the volcanic plume [8].

Lidar systems have increasingly been used since the early 1980s for the valuation of volcanic hazards (e.g., during the Soufriere and Mt. St. Helens eruptions in 1979 and 1981, respectively) [8]. Over the last 30 years, several lidars have been deployed to obtain data on the concentrations and fluxes of sulfate aerosol [27–29] and ash [30]. Furthermore, laser remote sensing systems have been used to detect volcanic particles in the troposphere at Mt. Etna (Italy) [31] and in the lower stratosphere during the Pinatubo eruption (Philippines) [32]. In particular, the lidar at Garmisch-Partenkirchen was one of the first examples of volcanic lidar implemented to observe the spread of the stratospheric volcanic plume (from 10 to 28 km) by evaluating the time-series of vertical profiles of the scattering ratio and vertically integrated column of the backscatter coefficient using a Nd:YAG laser operating at 532 nm. These measures were affected by the following average errors: peak scattering ratio: ±4–5% and integrated backscatter coefficient: ±10% [32].

DIAL-Lidar (Differential Absorption Light Detection and Ranging), or simply DIAL, is based on the fast switching of the wavelength of laser pulses on- and off-absorption of the molecule under study. From the application of the Beer–Lambert law to the ratio of the lidar signals (detected radiation vs. range) retrieved at both wavelengths, the range-resolved molecule concentrations (ppm) can be obtained, thus determining the two-dimensional or even three-dimensional plume morphology. This achievement is beyond the capabilities of FTIR, COSPEC (Correlation Spectrometer), and DOAS [15], which provide the path length integrated concentrations (ppm m). DIAL has been deployed at the main Italian volcanoes with UV lasers for tracking the total fluxes of SO_2 [28,33]. In particular, the first example of the quantification of volcanic SO_2 fluxes by using the DIAL (active) technique is reported in [33] where it was extensively compared with passive monitoring techniques (DOAS and COSPEC). In the previously mentioned work, the lidar-based SO_2 fluxes were 50% higher on average than those derived with other standard passive techniques (COSPEC and DOAS) commonly used in volcanology. This mismatch was interpreted as evidence of possible scattering-induced errors in the passive techniques.

More recently, ATLAS (Agile Tuner Lidar for Atmospheric Sensing), a ground-based DIAL system, has been implemented to perform measurements of water vapor flux at the Stromboli volcano (Italy) [34]. To our knowledge, this was the first time that lidar retrieved water vapor concentrations in a volcanic plume.

Finally, the Cloud-Aerosol Lidar with Orthogonal Polarization (CALIOP), an elastic backscatter lidar operated from a satellite, was used to observe volcanic ashes from 15 to 20 April 2010, just after the explosion of the Eyjafjallajökull volcano (Iceland) [35]. This has opened up new scenarios in future volcanic-lidar applications.

However, the apparatuses previously described were costly, heavy, and not effective enough to be implemented in the long-term period. Even though DIAL provides unprecedented opportunities to volcanologists, it requires more advancement to become a reliable technique for routine observations.

In order to fill these gaps, a new DIAL system, designed to measure the volcanic CO_2 flux, was developed as part of the ERC (European Research Council) Starting Grant Project "BRIDGE". The BrIdge voLcanic LIdar (BILLI), assembled at the ENEA (Italian National Agency for New Technologies, Energy and Sustainable Economic Development) Research Center of Frascati by the FSN-TECFIS-DIM group (Diagnostics and Metrology Laboratory) [36], successfully retrieved three-dimensional tomographies of volcanic CO_2 concentration in the plumes at Italy's most hazardous volcanic areas: Pozzuoli Solfatara (Naples, Italy) [21,37,38], the Stromboli volcano (Aeolians Islands, Italy) [39,40], and Mt. Etna (near Catania, Italy) [41,42].

As far as we know, lidar measurements of CO_2 have been performed in volcanic plumes only in recent times, due to the previously mentioned technical challenges and the extreme environmental operating conditions [14,21,37–42]. In the last decade, promising results have been obtained by using a variant of DIAL, known as the Integrated Path DIAL (IPDIAL), but mainly to monitor ambient CO_2 concentrations (e.g., power plants/industrial emissions), and an Open-Path laser Spectrometer (OPS) to measure volcanic CO_2 concentration and flux. However, despite the efforts to provide a remedy to some drawbacks of DIAL-Lidar systems (in terms of costs, complexity of the setup and maintenance, portability and integration time), the aforementioned systems had severe limitations such as the need of a topographic target to perform the measurements of CO_2, which were not range-resolved (only path averaged gas concentrations were retrieved). In addition, the volcanic CO_2 flux computed by OPS was affected by a large uncertainty [14]. Therefore, it can be concluded that BILLI was the first range-resolved DIAL system used to remotely measure both CO_2 concentration and flux in a volcanic plume.

In particular, as will be shown in the following paragraphs, BILLI has been deployed to retrieve range-resolved profiling in wide plumes, thus providing spatio-temporal dispersion concentration maps and high-resolution CO_2 flux time-series. It has opened unprecedented possibilities in measuring volcanic CO_2: BILLI made it possible to clearly detect an excess of a few tens of ppm over a distance of more than 4 km, considering a spatial and temporal resolution of 5 m and 10 s, respectively [42].

Quantifying the CO_2 output from Pozzuoli Solfatara (October 2014) [21,37,38] and Stromboli (June 2015) [39–41] allows for the interpretation, and if possible, the forecast, of the upcoming development of the volcano scenario, with huge benefits to the population living nearby. These examples show the huge social values of lidar systems in monitoring gas emissions. Moreover, observations at densely populated volcanic regions will have a high societal impact throughout Europe. In this regard, the potential of our system has also been confirmed by the results acquired during the last experimental campaign, carried out between July and August 2016 at Mt. Etna [42], retrieving a plume peak beyond 4 km.

The main goal of these experimental campaigns was to measure the exceedance of in-plume CO_2 concentration and flux to provide useful information to volcanological research for a precocious alert to the population in the event of eruptions.

In this paper, after a description of BILLI, special emphasis will be given to data processing and analysis of the experimental results, which will be thoroughly discussed. This was made possible by means of automatic software procedures written in MATLAB and based on the newly designed BRIDGE DIAL data processing technique. Significant CO_2 concentration profiles, dispersion maps, and time-series of CO_2 flux will be shown. Moreover, our monitoring was assessed by conventional data and local meteorological measurements.

Finally, thanks to the BILLI system and from a complete time-resolved plume evolution, it was also possible to retrieve the wind velocity (Stromboli field campaign, 2015), a useful parameter for the CO_2 flux calculation and for the mitigation of volcanic hazards, especially in the field of air traffic security.

2. Materials and Methods

2.1. The BILLI System

A lidar, also known as an optical radar, is an active range-resolving optical remote measurement system whose main components are composed of a transmitter (laser) and the receiver (telescope, optical analyzer/detector, data acquisition/computer). The principle of operation is the following [43,44]: the laser emits a light pulse into the atmosphere and its photons are scattered in all directions from molecules and aerosols (particles, droplets, etc.) present in the air. Part of the light is backscattered toward the telescope. Then, by means of a detector and computer, it is possible to analyze the detected signal vs. The time elapsed between the emission and detection (t). This makes it possible to characterize the chemico-physical properties of air along the beam. In fact, if R is the distance from the lidar of the backscattering layer and c is the speed of light, we can write R = ct/2. Going back and forth from the backscattering layer, the laser pulse undergoes attenuation by air, as expressed by its extinction coefficient, linked to the light absorption of its species. Often, the absorption lines of molecules are narrow, making it possible to employ DIAL to retrieve their concentration. This technique uses the idea of differential-absorption measurement in which two light pulses with slightly different wavelengths (ON: the light is strongly absorbed by the species under investigation, OFF: the light is not absorbed at all or at least much less) are emitted into the atmosphere, and two corresponding backscattered signals are detected concurrently. The difference between the two profiles is directly linked to the molecule abundance and this technique allows one to determine and map the concentrations of selected molecular species in ambient air [43,44].

BILLI (whose technical scheme and experimental setup are reported in Figure 1) is a monostatic-biaxial DIAL-Lidar system mounted in a small laboratory truck and is composed of a transmitter and receiver equipment [21,37–42].

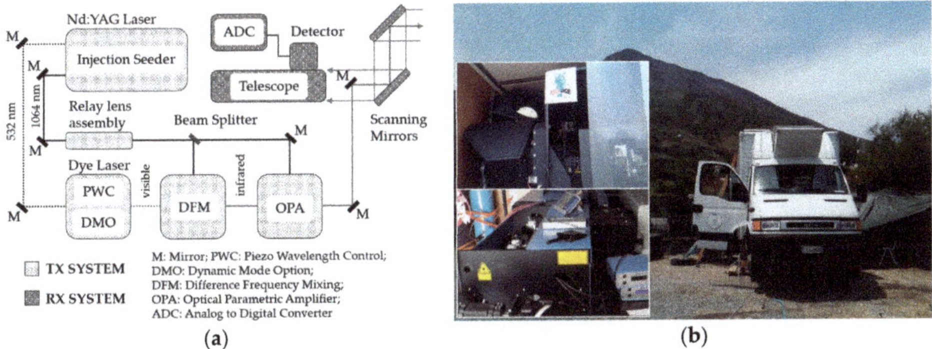

Figure 1. (a) Technical scheme of the BILLI DIAL system extrapolated and rearranged from the Figure 2 reported in [37]); (b) Experimental setup of the system during the Stromboli campaign (the volcano summit is visible above the two inset pictures). The scanning mirrors and the laser system are visible in the top and bottom inset picture, respectively.

The transmission sub-system is based on a double grating dye-laser optically pumped by an injection seeded Nd:YAG laser (powerful, tunable and narrow-linewidth), a device for the frequency generation (i.e., difference frequency mixing, DFM), and an optical parametric amplifier operating in the NIR (Near IR) band (OPANIR). The limited tunability (few tenths of nm) of Tm, Ho:YLF, and fiber lasers was the main reason of their exclusion (from the transmission sub-system during the implementation stage) given the difficulties in choosing the best absorption line [36]. Instead, due to the harsh environment near degassing craters that can cause burns on the surface of nonlinear crystals, OPOs (optical parametric oscillators) have been discarded [37–39].

This transmitter was used to generate laser radiation at ≈2 μm, a region of the electromagnetic spectrum absorbed by atmospheric CO_2, with negligible cross-sensitivity to H_2O [36].

The laser source of BILLI has the following characteristics [39]:

- 10 pm wavelength stability corresponding to about 0.2 cm^{-1} wavenumber stability;
- enhanced beam profile amplifier (EBPA);
- dynamic mode option (DMO); and
- piezo wavelength control (PWC).

The receiver sub-system is based on a Newton telescope and a detection system, which consists of a thermoelectric-cooled InGaAs PIN photodiode, directly linked to the analog-to-digital converter (ADC). Two large motorized elliptical mirrors (major axis: 450 mm) make it possible to scan all of the angles above the horizon. The specifications of the whole system are reported in Table 1, and more details are available in previous works [21,37–42].

Table 1. Main specifications of the BILLI system during the BRIDGE field campaigns.

Transmitter	Pulse energy	25 mJ
	Pulse duration	8 ns
	Repetition rate	10 Hz
	Wavelengths	ON: 2009.537 nm, OFF: 2008.484 nm *(Stromboli Volcano 2015, Mt. Etna 2016)*
		ON: 2012.194 nm, OFF: 2011.1 nm *(Pozzuoli Solfatara 2014)*
	Laser linewidth	0.04 cm^{-1}
	Beam divergence	0.5 mrad
Receiver	Mirror coating	Al
	Clear aperture	300 mm
	Focal length	900 mm
	Field of View	1 mrad
Scanning elliptical mirrors	Mirror coating	Al
	Clear aperture	320 mm × 451 mm
Detector module	Photodiode	InGaAs PIN
	Diameter	1 mm
	Responsivity	1.2 A·W^{-1}
	Gain	5.1·104 V·A^{-1}
	NEP	10 pW·Hz$^{-1/2}$
	Bandwidth	0–10 MHz
Analog-to-digital converter (ADC)	Dynamic range	14 bit
	Sampling rate	100 MS·s^{-1}

From Table 1, it is possible to note that, with respect to the Pozzuoli Solfatara field campaign [21,37,38], the operative wavelengths were slightly changed during the campaign carried out at Stromboli and Mt. Etna [39–42]. This was due to the different mode of operation of the system.

In fact, in the first campaign, the line (2012.194 nm) was selected looking for low cross-sensitivity to water vapor and high absorption coefficient [37,38] that was optimum for short-range measurements of CO_2 concentration, whereas, in the other cases, the main goal was to reach longer ranges. To do this, the differential absorption cross section of carbon dioxide ($\Delta\sigma$) was reduced and a different ON line (2009.537 nm) was chosen [39,42]. For more details, see Table 2 [21,37–42].

As will be shown in the following, it was observed that the measurement uncertainty was governed by instability in the wavelength set. For this reason, from the Stromboli campaign, a photo-acoustic cell—with CO_2 at atmospheric pressure and temperature—was mounted at the laser output, thus providing wavelength accuracy, repeatability, and stability, as has been shown in the laboratory [36–42].

Table 2. Wavelengths, wavenumbers, and CO_2 absorption coefficients calculated (at T = 296 K, P = 1 atm) during the field campaigns of Pozzuoli Solfatara, Stromboli volcano, and Mt. Etna.

	Pozzuoli Solfatara, October 2014		Stromboli Volcano, June 2015 and Mt. Etna, July 2016	
Mode	ON	OFF	ON	OFF
Wavelength [nm]	2012.194	2011.1	2009.537	2008.484
Wavenumber $[cm^{-1}]$	4969.7	4972.403	4976.271	4978.880
CO_2 Abs. Coeff. $[m^{-1}]$	7.474	0.331	1.934	0.302
CO_2 $\Delta\sigma$ $[m^{-1}]$	7.143		1.632	

Moreover, a new version of the software was developed: the step motor that modifies the resonator cavity inside the dye laser now allows one to shift the emitted wavelength, and the photo-acoustic signal was measured, averaging for 1 s (i.e., for 10 laser shots). In this way, the wavelength shown by the laser monitor can be corrected for a small shift (0.12 cm^{-1}) and the experiment and theory [45] agreed within the linewidth of the emitted pulse (0.04 cm^{-1}). This leads to being able to control the wavelength with the photo-acoustic cell before every lidar measurement, so that the operator can fix the ON and OFF wavelengths with a smaller error than the emitted linewidth [39].

Therefore, a gradual refinement of the system performances was obtained in the last two in-field campaigns, thanks to the following measures:

- an appropriate choice of CO_2 absorption cross section, which enabled both short- and long-range measurements;
- the implementation of the photo-acoustic cell, resulting in a containment of systematic uncertainty associated with measurements of CO_2 concentration (see Section 2.2.5); and
- the reduction of the laser beam divergence due to a better optical setting, resulting in an improvement in the signal-to-noise ratio (SNR).

As will be shown in the following, this allowed us to detect CO_2 plume peaks beyond 4 km and to improve the accuracy of the CO_2 flux measurements. Finally, efforts have been spent to improve the system portability and, in particular, to integrate existing data processing routines [42].

Summarizing, the main characteristics of the system are as follows:

- the lidar is able to explore the troposphere in both vertical and horizontal directions;
- every lidar return is obtained averaging at least 50 shots ON and OFF (interlaced between them with $t_{shift} = 0.1$ s);
- the temporal resolution of the lidar echoes (Δt) is about 10 ns, corresponding to a ΔR of 1.5 m, although the actual time resolution of BILLI is 5 m: this latter value corresponds to the time response of the detection system and is linked to its bandwidth [39]);
- a concentration profile can be obtained in about 10 s by a couple of lidar signals (ON and OFF) using the newly designed mathematical technique explicitly developed for this application;
- starting from lidar signals of the same scan, it is possible to retrieve the dispersion map of in-plume CO_2 concentration (ppm) in the investigated area;
- knowing the (estimated) plume velocity, it is possible to obtain the carbon dioxide flux ($Kg·s^{-1}$) [21,41,42]; and
- starting from lidar profiles acquired successively, it is possible to track the motion of discrete atmospheric particles emitted by the volcanic crater, and this allows one to also estimate the wind velocity [38] (more details are reported in Section 3.2).

However, a comprehensive overview of the special features of BILLI has been reported in previous works [21,37–42].

2.2. The BRIDGE DIAL Data Analysis Technique

As above-mentioned, from the literature [44], it is well known that the radiation returned to the lidar telescope vs. time is linked to the backscattering of the laser pulse by an atmospheric layer at a specific range from the system. Consequently, lidars provide range-resolved characterization of the atmosphere and of its components (particulates and molecules) along the path of the laser pulse. This information can be displayed by means of intensity vs. distance graphs.

Considering that for our experiments a DIAL system was deployed, two different wavelengths—λ_{ON} and λ_{OFF}, absorbed and not absorbed, respectively, by carbon dioxide—were used to acquire the intensity profiles. The difference between the two wavelengths was very small and, as a consequence, the behavior of the atmosphere (e.g., considering aerosol and molecular attenuation and turbulence effect on the beam transmission) was the same, except from the CO_2 absorption. Thanks to the difference in return between the nearly simultaneous λ_{ON} and λ_{OFF} signals (as stated in Table 1, the time interval between two successive laser pulses was 0.1 s), and using the previous mentioned technique, the operator can retrieve the range-resolved CO_2 concentration profile in volcanic plumes.

The BRIDGE DIAL technique is a mathematical procedure written with MATLAB software that has been specifically developed for real-time data processing and graphic presentation of data, acquired during the field campaigns carried out by the BILLI system. As recalled before, throughout the years, both the system and the data analysis technique have been optimized. In this section., only the main steps of the technique are reported, and more details are given in previous works [21,37–42].

2.2.1. Preliminary Steps

Generally, raw data are: (i) normalized by dividing the lidar signal by the laser energy; and (ii) background-noise subtracted [44]. Background noise here is measured by averaging the farthest part of the lidar echo (see Figure 2a(1)).

To simplify, during its travel to distance R, the pulse energy decreases: (i) exponentially, due to air extinction; and as $1/R^2$, simply because the detected photons are directly proportional to A/R^2 (i.e., the solid angle of the telescope), where A is its effective area. To compensate for this effect, the lidar signal P(R) (where P is the received power as a function of range R) is commonly transformed into the range-corrected signal (RCS). This is performed by multiplying the signal P(R) by R^2 [44] (see Figure 2a(2)).

Finally, to further improve the SNR (Signal-to-noise Ratio), up to 100 laser shots have been averaged for every lidar return and a Savitzky–Golay filter [46] algorithm with 13 points was used (see Figure 2a(2)).

2.2.2. Conversion of RCS (Range Corrected Signal) Profiles Into Range-Resolved Profiles of in-Plume Excess CO_2 Concentrations

The core of the BRIDGE DIAL technique consists of the following steps [21,37–42]:

- Evaluation of CO_2 concentration in the natural background atmosphere (C_0) outside the plume [39].
- Evaluation of the average excess CO_2 concentration (ΔC) in the volcanic plume cross-section (e.g., between two rockfaces as shown in Figure 1 [39]) encountered along the laser optical path in the examined region with respect to C_0.
- Calculation of the range-resolved excess of in-plume CO_2 concentration, corresponding to each i-th ADC channel of the lidar profile from (see Figure 2b):

$$C_{CO_2,i} = k \cdot RCS_i \tag{1}$$

$$k = \frac{\Delta C (R_1 - R_2)}{\Delta R \sum_i RCS_i} \tag{2}$$

where k is the multiplication factor; ΔC—as we mentioned—is the average excess carbon dioxide concentration measured between the two rockfaces and is inferred from the ratio of the corresponding two lidar echoes; R_1 and R_2 indicate the ranges of the two rockfaces; ΔR is the range interval corresponding to an ADC channel; and RCS is the range corrected signal.

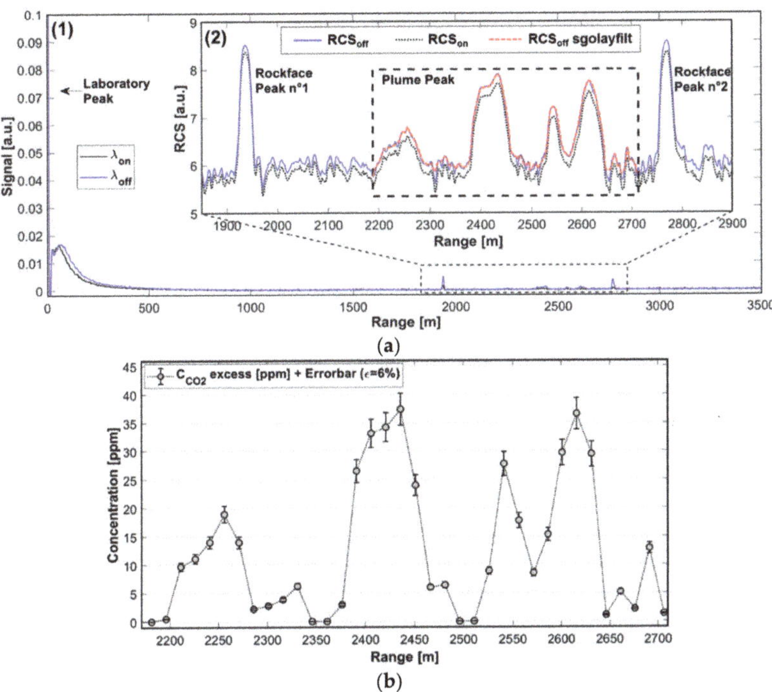

Figure 2. (**a**) Example of a lidar profile acquired during the Stromboli volcano field campaign on 26 June 2015. (1) Data profiles in the form of signal intensity (arbitrary units, a.u.) vs. range (in meters) cleaned by the noise. (2) Detail of the experimental RCS (Range Corrected Signal) and filtering of the OFF channel using a Savitzky–Golay filter algorithm with 13 points in which the plume peak is visible between two rockfaces in the range interval between 1800 and 2900 m. The laboratory peak in (1) provides the start time of the pulse emission and is due to the reflection of some laser radiation inside the mobile laboratory. The two rockface peaks (2), as their name indicates, are due to pulse backscattering by the rockfaces of Stromboli. Finally, the peaks from 2200 to 2700 m are linked to the backscattering of the laser pulse from the plume. (**b**) Particulars of the exceedance of in-plume CO_2 concentration profile (acquired at 11:27, with an elevation 16.98° and azimuth 237.8°). The quadratic sum of statistical (2%) and systematic (5.5%) uncertainties led us to estimate an overall uncertainty of about 6%.

It should be noted that, during the in-field preliminary setup, the system was optimized to perform measurements in selected directions, where one or more topographic targets were partially inside the lidar FOV (Field of View). This allowed us to retrieve some useful parameters (see C_0, ΔC, and k) for the initialization of the data analysis technique. Once the multiplication factor is known, it is possible to perform sequential measurements multiplying each range-corrected lidar profile by this factor, within the limits of the expected dispersal area of volcanic plume.

- Drawing CO_2 dispersion maps by scanning the selected region and arranging subsequent concentration profiles (acquired at different scanning angles and relative to the same scan) in chronological order.

2.2.3. Assessment of the Plume Transport Velocity

The plume velocity v_p (m·s^{-1}) can be inferred from UV camera photograms [21,41,42] or assessed by implementing the correlation method [40,47].

The latter case consists of the following main steps [40]:

- Firstly, RCS is normalized by the laser emitted energy on a shot-per-shot basis (see Table 1) to define a normalized signal $S(R_i)$ independent of the energy fluctuation. Then, the mean normalized signal $\bar{S}(R_i)$ is obtained by averaging N successively recorded individual signals at successive times t_n.

- Second, once we know $S(R_i)$ and $\bar{S}(R_i)$, it is possible to retrieve the information about the shot-per-shot fluctuations of the signals, with respect to the mean normalized signal that is contained in the fractional deviations (specific datasets in which the time evolution of in-plume spatial aerosols inhomogeneities is clear) as:

$$f(R_i,\, t_n) \;=\; \frac{S(R_i, t_n) - \bar{S}(R_i)}{\bar{S}(R_i)} \tag{3}$$

- Third, to determine the displacement of the backscattering inhomogeneities, a cross-correlation function $\rho(L,\, \Delta H)$ is defined with respect to an altitude interval ΔH and to a lag value L, corresponding to a discrete number of spatial points as:

$$\rho(L,\, \Delta H) = \frac{1}{\sigma \sigma_L} \sum_{i=a}^{b} \sum_{n=1}^{N-1} f(R_i,\, t_n) f(R_{i+L},\, t_{n+1}) \tag{4}$$

and a superimposed Gaussian estimator is employed on the data to calculate the mean and standard deviation values. These values, respectively, are used to infer the radial wind velocity and its uncertainty.

- Finally, it is possible to measure the horizontal component of the wind velocity (in this kind of application, the vertical component of wind can be neglected) as:

$$v = \frac{v_r \cos \theta}{\cos \Delta \Phi} \tag{5}$$

where v_r is the radial wind in the altitude interval ΔH determined from the lag value for which the correlation is maximum; θ is the elevation angle of the telescope; and $\Delta \Phi$ is the difference in the azimuthal orientation of the wind vector and of the light path.

It should be noted that the described experimental technique (for the tracking of wind velocity) has the advantage of being performed remotely, rapidly, and without additional costs, thanks only to a further data processing routine integrated in the main data analysis software. Therefore, using the same dataset, BILLI is able to perform multiple measures of in-plume CO_2 concentration (in excess), flux, and simultaneously, a good estimation of the plume transport velocity over large areas. This can also be useful to provide a fast tracking of volcanic ash, in order to improve air traffic security. Nevertheless, although promising results (in line with in-situ weather stations) have been obtained using the correlation method, the main drawbacks are still high measurement uncertainty and discrepancy with respect to other optical devices (e.g., UV camera images), probably due to the high fluctuations of wind as a function of the altitude and the differences between the devices and/or data acquisition/processing techniques [40].

2.2.4. Evaluation of CO_2 Flux

The values of CO_2 concentrations reported both in the profiles (see Figure 2b) and the dispersion maps (see Figure 3a, Figure 4a, Figure 5b) were used to compute the carbon dioxide flux (see Figure 3b, Figure 4b, Figure 5c). To this aim, it is necessary to integrate the excess (i.e., background corrected) carbon dioxide concentrations over the whole cross-section of the plume determined by each scan, and multiply this amount by the transport velocity of the plume. Formally, the carbon dioxide flux (Φ_{CO_2}, in $Kg·s^{-1}$) is calculated from the following equation [21,41,42]:

$$\Phi_{CO_2} = v_p·\frac{PM_{CO_2}}{10^3 N_A}·N_{molCO_2-total} \tag{6}$$

where v_p is the plume transport velocity ($m·s^{-1}$); $N_{molCO_2-total}$ is the total plume carbon dioxide concentration ($molecules·m^{-1}$); and PM_{CO_2} and N_A are the CO_2 molecular weight and the Avogadro's constant, respectively. $N_{molCO_2-total}$ was calculated by integrating the effective average excess CO_2 concentrations ($\overline{C_{exc,i}}$ in ppm) over the whole plume cross section, according to:

$$N_{molCO_2-total} = N_h·10^{-6}·\sum_i \overline{C_{exc,i}}·A_i \tag{7}$$

where N_h is the atmospheric number density [$molecules·m^{-3}$] at the crater; 10^{-6} converts it into parts per million (ppm); and A_i represents the i-th effective plume area provided by:

$$A_i = l_i·\Delta R \tag{8}$$

where ΔR is the ADC resolution (1.5 m) and l_i is the i-th arc of the circumference:

$$l_i = R_i·\theta \tag{9}$$

where R_i is the i-th range vector (m) and θ is the angle resolution (rad).

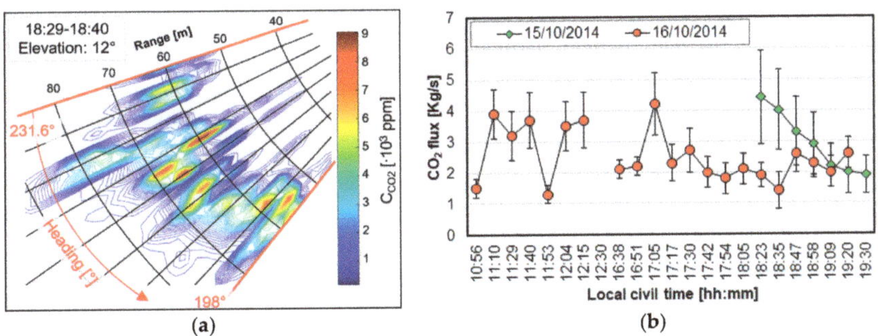

Figure 3. (a) An example of a lidar scan in polar coordinates through the plume at an elevation of 12° (15 October 15 2014). The contour lines show isopleths of background-corrected CO_2 mixing ratios in the plume (expressed in ppm, the legend is the vertical colored bar), shown as a function of heading and range. (b) Time-series of lidar-derived CO_2 fluxes from Pisciarelli obtained from 15 October (afternoon) to 16 October (afternoon), rearranged from Figure 4 reported in [21]. The timescale for the morning scans on 15 October is shown in green. Each point refers to a particular scan through the plume (time is the onset time of the scan, with each scan lasting 10–20 min) [21].

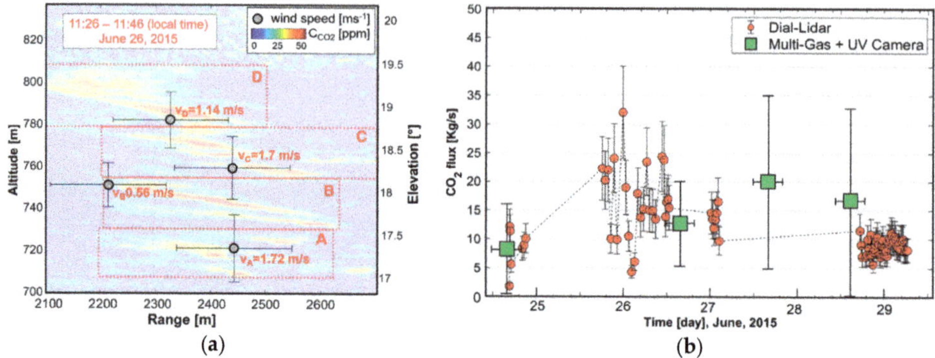

Figure 4. (**a**) In-plume CO_2 concentration map acquired on 26 June 2015 at Stromboli. Carbon dioxide concentration varies as a function of increasing altitude (on the left of the map) and elevation (on the right of the map). The superimposed grey points represent the vertical wind profile retrieved by BILLI in the same dataset. Here, the mean value of the wind velocity was found about 1.3 m/s. The x-axis error bar is the weighted average standard uncertainty of wind velocity, equal to 0.52 m/s. Instead, the error bar on the y-axis represents the altitude halfwidth between adjacent regions (A, B, C, and D) [40]). This figure has been obtained by combining Figure 4 and Figure 8 reported in [40]. (**b**) Time-series of CO_2 fluxes from Stromboli volcano from 24 to 29 June 2015, rearranged from Figure 5 reported in [41]. The BILLI based fluxes (red circles) were obtained using the procedure detailed here [39,41]. For comparison, independent CO_2 flux estimates, obtained by multiplying the in-plume CO_2/SO_2 ratio (from Multi-GAS) by the SO_2 flux (from UV cameras) are also presented (green squares) [41].

2.2.5. Evaluation of Uncertainty Sources

The main sources of uncertainty that affected our CO_2 flux measurements are reported below [39,41]:

1. Systematic uncertainty of the carbon dioxide concentration measurement. Taking into account the experience [21,41], the systematic uncertainty of the BILLI system was dominated by inaccuracies in wavelength setting [39], and this led to errors in differential absorption cross section and, as a consequence, in CO_2 mixing ratio. In this regard, in order to control the transmitted wavelength before each atmospheric measurement, a photo-acoustic cell filled with pure CO_2 (at atmospheric pressure and temperature) was implemented close to the laser exit. This allowed setting the ON/OFF wavelengths used in this study with an accuracy better than the laser linewidth [39], with an overall uncertainty in the wavelength setting equal to ± 0.02 cm^{-1} (half laser linewidth). For these reasons, the systematic uncertainty of the CO_2 concentration measurement was below 5.5% [41,42].

2. Statistical uncertainty of the carbon dioxide concentration measurement. Starting from the calculation of the standard deviation of the signals corresponding to every ADC channel, and implementing the standard uncertainty propagation technique [48], it has been possible to evaluate the statistical uncertainty. This uncertainty varies as a function of distances from the system, considering typical atmospheric and plume conditions encountered during field campaigns. Typical values at 2.5 km and over 4 km from the system were estimated to be equal to 2% and slightly above 5%, respectively [41,42].

3. Uncertainty linked to plume transport velocity. Mean and standard deviation of the wind velocity were calculated, also in this case by the standard uncertainty propagation technique. Then, the corresponding relative uncertainty was evaluated [48]. The assessed values ranged from 3% (at Stromboli [41]) to 19% (at Pozzuoli [21]).

4. Uncertainty in the characterization of the integration area. Due to difficulties in accurately assessing the surface where there is an excess of carbon dioxide concentration, this parameter

represented the main source of uncertainty for the calculation of volcanic CO_2 fluxes, with an order of magnitude estimated to be about 25% [41].

If all uncertainty sources are statistically independent, all the components have to be quadratically summed [48] in order to obtain the total uncertainty of volcanic CO_2 fluxes, which varies from 20% to 30% (dominated by the area uncertainty [41,42]). This value is consistent with the results obtained during the field campaigns of Pozzuoli (2014) [21], Stromboli (2015) [41], and Mt. Etna (2016) [42].

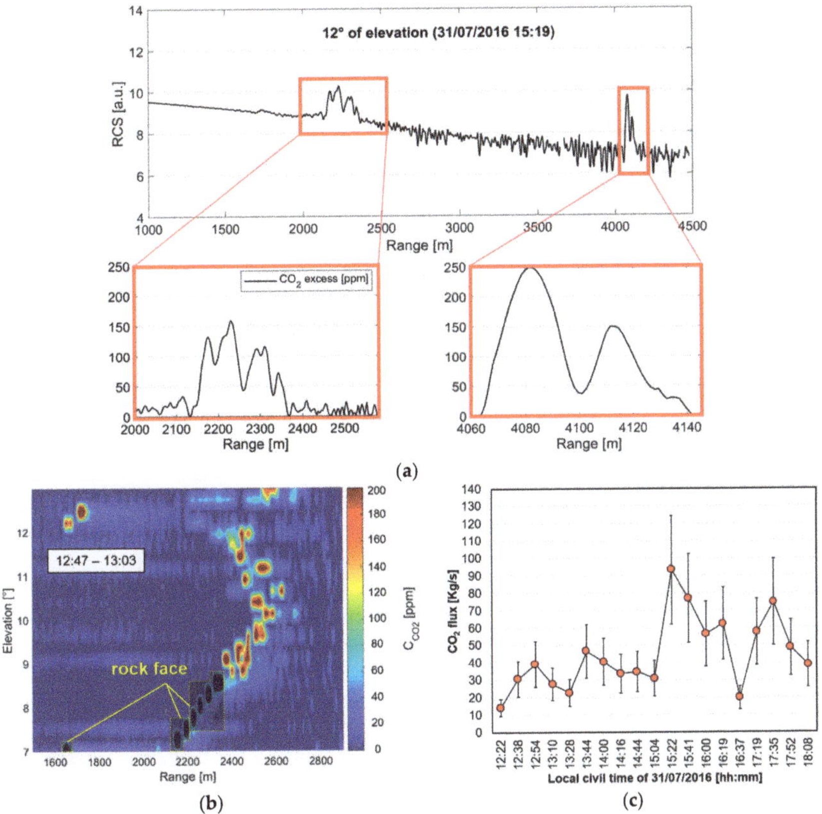

Figure 5. (**a**) Lidar return at 12° of elevation: two wide and jagged peaks from the volcanic plume are clearly visible; the CO_2 profiles inside the volcanic plume are shown in the two red boxes. The uncertainty was equal, respectively, to 6% for the left-side plume peak and 7.5% for the right-side plume peak. (**b**) Vertical scan (fixed azimuth: 230°) of the volcanic plume (CO_2 excess) acquired on 31 July 2016 (12:47 p.m.–1:03 p.m., local civil time). With that fixed azimuth and in the elevation range 7–9°, rockfaces reflect the laser pulse, corresponding to peaks that are not linked to the plume (rockfaces are disentangled from plume because of the narrowness of peaks). (**c**) Carbon dioxide flux measured at the northeast crater on 31 July 2016 (12:22 p.m.–6:08 p.m., local civil time). The bars show the error in carbon dioxide flux (33%). Please note that the previous figures were extrapolated and rearranged from Figure 2a, Figure 4c, Figure 5, respectively, reported in [42].

3. Results and Discussion

A brief overview of the experimental campaigns carried out over three years by the BILLI system will be presented and discussed here.

The first experimental campaign was carried out during October 2014. The system was placed $\approx$100 m far from the fumaroles of Pozzuoli Solfatara (Pisciarelli, Phlegraen Fields). The system performed a horizontal scan at different elevations, retrieving, for the first time, CO_2 concentration dispersion maps and time-series of lidar-derived CO_2 fluxes [21,37,38].

About ten months later (June, 2015), other tests were carried out at Stromboli Volcano Island, where BILLI operated nearly 24 h for almost a week, performing horizontal and vertical scans. CO_2 excess of a few tens of ppm was measured in some minutes in the volcanic plume up to a distance of nearly 3 km far from the system. In addition, in this case, daily cycles of CO_2 flux were successfully retrieved [39–41].

Finally, the potential of the BILLI DIAL system was confirmed by the last field campaign at Mt. Etna at the end of July 2016, where, for the first time, in-plume CO_2 concentration peaks were retrieved at a distance exceeding 4 km [42].

It should be noted that during all field campaigns, eye safety regulations were followed and weather did not alter the BILLI operation, except for a moderate influence of the wind, as is reported in Section 3.2 [21,37–42].

3.1. The Experimental Campaign of Pozzuoli Solfatara

The first field experiment was conducted at the Phlegraen Fields volcano [21,37,38]; this is a rising caldera including Pozzuoli and the suburbs of Naples, among the most inhabited zones of Italy. The bradyseismic crises of the volcano produced thousands of shallow earthquakes, and its activity has been often, even if not continuously, active in the last few decades. Scientists and stakeholders observed these recent crises, despite them not evolving into an eruption, bearing in mind the related important hazard mitigation problems and the linked necessary interventions. These issues have become even more critical taking into account that ground uplift restarted in 2012, with the intensification of degassing activity, in the caldera center and at the main surface hydrothermal manifestations of Solfatara and Pisciarelli. Current uplift/degassing unrests observed in Pozzuoli are probably due to intermittent injections of carbon dioxide-rich gas in the hydrothermal system of the volcano, indicating that fresh magma is supporting a strong inflow of gas from deep layers. Here, attention has been focused on the degassing fumarolic vent of Pisciarelli that, besides a small augmentation in discharge temperature (from less than 95 °C up to 110 °C), has visibly escalated in degassing activity over the past decade (the carbon dioxide concentration in fumaroles has also risen), making it the most active among the Italian fumarolic vent areas. Therefore, quantifying the CO_2 output from Pisciarelli is vital for the interpretation and possibly the prediction of future changes in this volcanic zone [21].

The system measured nearly continuously during both 15 and 16 October 2014, after the experimental setup on 14 October. During the measurement operations, BILLI performed a horizontal scan of the almost vertical plume in the 198°–232° range of azimuth angles (see Figure 3a). The elevation of 12° was used mostly for the performed scans, where the plume structure was fully resolved and the lidar echoes were more intense. Controls proved that at lower and higher elevations, the plume was often either optically too thick or too spread, so only a few additional scans at 0°, 6°, and 15° were completed. A total of 200 lidar returns (100 at λ_{ON} and 100 at λ_{OFF}, see Table 2) were ascribed to a single profile and averaged to improve the SNR; in this way, the sampling frequency of the lidar echoes was 0.05 Hz, which corresponded to a 20-s time resolution. As already discussed, the space resolution of BILLI was 5 m. The combination of about 25 profiles yielded a total scan of the plume in around 10 min [21,37,38].

Thanks to the BRIDGE DIAL technique (see Section 2.2) it was possible to retrieve, globally, up to 40 scans of CO_2 concentration in excess in the plume and, at the same time, the CO_2 flux time-series. The whole dataset covered almost three days of operation, for a maximum of nine consecutive hours. Examples of the CO_2 dispersion map and flux time-series are reported in Figure 3a,b, respectively [21].

Figure 3a illustrates the results of a typical scan (using the polar diagram) through the plume. When the laser beam intercepted the plume, peak CO_2 concentrations of up to 5000 ppm were measured.

The plume was typically crossed at azimuth angles of 205–225° and at distances in the range of 50–70 m from the system. At azimuth angles <205° and >225°, or at distances <50 and >70 m, BILLI retrieved the carbon dioxide concentration usually found in ambient air, thus providing confirmation that the laser beam was able to scan the whole plume. This result directly corresponded to the most vigorously degassing fumarolic vent of Pisciarelli. An evident CO_2 peak, in scans performed at an elevation angle of 6°, was even clearer in the same fumarolic vent. This fact was also due to the reflection of the laser beam off the rockfaces enclosing the vent [21,37,38].

Figure 3b, instead, illustrates CO_2 flux time-series from the fumarolic system. Here, the cross-correlation analysis, applied to sequences of visible images of the moving plume that were co-acquired at a high rate (30 Hz) by a digital optical camera, was applied to achieve the vertical plume transport velocity. The overall error in the calculated plume velocity was in the range from 10 to 28%, leading to a total error of CO_2 fluxes from 15 to 33% [21].

Thanks to the high time resolution (around 10 min) of the system, it was possible to follow important intraday dynamics, nearly impossible to measure with conventional instruments for several reasons such as environmental difficulties, high costs, instrumentation ineffectiveness, and the larger amount of ambient air carbon dioxide that is present along the line of sight between the spectrometer, target (plume), and the source of photons (sun, magma, and lava fragments).

The strong changes over timescales of tens of minutes are clear in the Pisciarelli hydrothermal site such as from a peak emission of 4.4 Kg/s (at 18:23 h local civil time on 15 October) down to 1.9 Kg/s after about 60 min. Apart from these fast variations, our data indicate a general constancy of CO_2 output over timescales of a few days. The mean carbon dioxide emissions for 15 and 16 October 2014 were quite similar (2.63 ± 0.98 and 2.52 ± 0.84 Kg/s), corresponding to the total daily outputs of 256 ± 89 and 218 ± 71 tons, in that order. Furthermore, the carbon dioxide fluxes measured by BILLI were in overall agreement with other estimations carried out with in situ data obtained with a portable gas analyzer [21].

Nevertheless, our system has the following advantages: (i) external light sources are not required; (ii) measurements can be performed without risk from a remote area; (iii) the spatio-temporal resolution is finer; and (iv) the morphology of the plume can be obtained in some minutes.

Moreover, the big difference between the lidar echoes coming from the plume and ambient air allows the operator to disentangle the carbon dioxide of volcanic origin, and therefore to quantify the volcanic carbon dioxide flux. Such measurements are new in geophysics, and promise to contribute to volcanological research.

Even though, at this experimental stage, a considerable number of further developments are needed (e.g., system miniaturization, automation, independent measurements of in-plume wind velocity, and long-distance remote sensing), the system performed the first measure of CO_2 concentration and flux in a volcanic plume [21].

3.2. The Experimental Campaign of Stromboli Volcano

The second field experiment was conducted at Stromboli Volcano Island. The volcanic activity on the island is generally considered "regular" and mild (strombolian). Larger-scale vulcanian-style explosions occasionally occur. Although these events, locally referred to as "major explosions" or "paroxysms", are not frequent and are short-lived (from tens of seconds to a few minutes), they could be extremely dangerous for locals and authorized personnel, since they produce fallout of coarse pyroclastic materials over wide dispersal areas. Moreover, neither geophysical nor volcanological precursor signals were observed before these events occurred. This could be due to the fact that they originate deep in the crustal roots of the volcano plumbing system. However, it has been observed that the previously cited events are systematically preceded by days/weeks of anomalous CO_2-rich gas leakage from the Stromboli's deep (8–10 km) magma storage zone. For these reasons, volcanologists are considering with increasing interest CO_2 flux emissions from the open-vent crater plume for the monitoring of volcanic activity and hence for the assessment and mitigation of the related hazards [41].

As in other volcanic areas, at Stromboli, the volcanic gas CO_2 from a combination of simultaneously measured SO_2 fluxes and plume CO_2/SO_2 ratios has also been recently evaluated [7,11]. Even though the remote measure of SO_2 flux is possible by implementing UV spectroscopy [16], the CO_2/SO_2 ratio requires in situ direct sampling/measurements close to hazardous active vents. For the previous reasons, the implementation of the BILLI DIAL system for the remote observation of the volcanic CO_2 flux, from distant and less risky locations, is a real advantage.

The system operated almost continuously from 24 to 29 June 2015 despite the harsh environmental conditions due to the presence of high humidity, acid vapors, and re-suspended dust. It was placed inside the fence of the local ENEL (main Italian electric company) power plant in the Scari area, ≈ 2.5 km from the degassing vents on the volcano summit, between the margins of morphological peaks known as Pizzo and Vancori, 918 and 924 m, respectively, above sea level (ASL). The laser beam typically scanned the azimuth angles from 235.3° to 253.6° and the elevation angles from 15.2° to 27.4°.

As stated in Section 2.1, in order to reduce the differential absorption cross section of CO_2, the transmitted wavelengths were changed with respect to the previous campaign (see Table 2). As a result, longer distances were reached by the laser beam. Furthermore, the implementation of a photo-acoustic cell filled with pure CO_2 allowed us to reduce the systematic uncertainty, dominated by instability in wavelength setting, to 5.5% (see Section 2.2.5). Compared to the previous experimental campaign, the pulse repetition rate remained unchanged (10 Hz), instead both the temporal resolution was slightly modified and set at 10 s because only 100 lidar returns (50 at λ_{ON} and 50 at λ_{OFF}, interlaced) were averaged for each scan to increase the SNR. A further improvement of the SNR and, therefore, of the accuracy of both CO_2 concentration and flux measurements was possible by employing filtering algorithms and curve fitting techniques, in the preliminary data analysis steps (see Section 2.2.1). Furthermore, by combining about 50 profiles, it was possible to obtain both vertical and horizontal plume scans in less than 10 min. Eventually, by processing several repeated scans (usually 10) acquired at different elevations, we also retrieved three-dimensional tomographies of the volcanic plume [39–41].

Thanks to an improved and fully automated version of the BRIDGE DIAL data processing routine (taking into account local conditions and the changes in system setup), it was possible to retrieve up to 180 scans of CO_2 concentration in excess in the plume and, at the same time, the CO_2 flux time-series. In addition, thanks to a further data processing routine (see Section 2.2.3), it was possible to estimate the plume transport velocity and, hence, the local wind velocity. The whole dataset covered almost six days of operation, for a maximum of 12 consecutive hours. Examples of CO_2 dispersion map, wind profiles, and flux time-series are reported in Figure 4a,b, respectively [39–41].

Figure 4a illustrates the results of a typical vertical scan through the plume as a function of range (X axis), elevation angle, and altitude (Y axes). The tomography reported above was extrapolated from the measurement session performed on 26 June 2015 between 11:26 and 11:46 (local civil time), and it represents a dataset acquired as a single group of measurements slowly stepping up in elevation angle (from 16.8° to 20°) over the course of about 20 min, since, during that time, the plume displacement was extremely clear. The carbon dioxide concentration in the zone under study is indicated by the color scale, with peaks exceeding 60 ppm. Here, the volcanic plume was encountered in the 2150–2600 m range, from 17 to 19.5° elevation angles, and a fixed azimuth of 237.8°. Regions highlighted with red boxes and named as A, B, C, and D were posited to be the same set of scatterers being tracked over time. Therefore, they represent a useful example for summarizing the time evolution of inhomogeneities relative to the volcanic plume. To appreciate these displacements, it is necessary to specify that particles detected in region A need a certain amount of time to move to the regions B (about 3 min), C (about 6 min), and D (about 9 min). As observed in the field, these movements were mainly due to the presence of wind blowing from north–northwest [40].

Considering both vertical and horizontal scans of the whole campaign, the beam hit the plume at azimuth angles in the range 237–251°, elevation angles in the range 16.5–20° (which means altitude of 700–830 m) and at distances in the range 2200–2500 m. For azimuth angles <237° and >251°, or for distances <2200 and >2500 m, BILLI retrieved the carbon dioxide concentration we usually find in ambient air, thus providing confirmation that the laser beam was able to scan the whole plume. This result was confirmed by direct observations [37,39].

The quality of the results (as in the example shown in Figure 4a), combined with the high temporal resolution of the acquired data, allowed us to perform, almost simultaneously with in-plume CO_2 concentrations, the first independent measurement of plume transport velocity (v_p) and therefore of the wind velocity, using the correlation method mentioned in Section 2.2.3 and detailed in a previous work [40]. As already mentioned, the assessment of the wind velocity could be crucial for both the CO_2 flux retrieval and air traffic control applications. In this regard, a set of data (lasting about 90 min) was selected from the morning of 26 June 2015.

An example of vertical wind profile was superimposed on the same dataset (lasting about 20 min) of the CO_2 concentration dispersion map reported in Figure 4a, showing a high consistency with the time evolution of the in-plume spatial aerosol inhomogeneities (see the previously mentioned A, B, C, D regions highlighted with red boxes).

The global weighted average value of wind velocity was found equal to 1.1 ± 0.4 m/s for the whole measurement test session. Although promising in terms of cost saving and speed, our measurements were slightly different from the ones acquired by the UV camera [41] and affected by a high uncertainty level due to the presence of wind blowing in the perpendicular direction, with respect to the system location, the worst observing condition for the tracking of the wind [40]. Nonetheless, they were found in agreement, within the error bar, with weather data acquired by conventional devices and with a precision comparable to that of stand-alone fixed-beam correlation lidars. It should be noted that, a comprehensive comparison with local data was considered unreliable, since the nearest weather station on the island performed 6-h average of weather parameters acquired at 4 m ASL and not at the same height of the laser beam pointing. For these reasons, efforts will have to be devoted to advance the configuration of BILLI (e.g., implementing at least three beams or a 360-degree scanning mirror) to fully and accurately retrieve the three components (x, y, and z) of the wind velocity [40]. Therefore, due to the constraints (in terms of dataset duration and results reliability) of our preliminary tests on the fast tracking of wind velocity, during the Stromboli campaign, CO_2 flux time-series were retrieved using v_p values obtained from processing the sequences of UV camera images [41].

Once having obtained the CO_2 dispersion maps and the plume transport velocity using the method described in Section 2.2.4, it was possible to retrieve the CO_2 flux time-series with a cumulative uncertainty equal to 25% (dominated by the integration area uncertainty), as reported in Figure 4b [41]. The results confined the CO_2 flux during 24–29 June 2015 in the range from 1.8 ± 0.5 to 32.1 ± 8.0 Kg/s. The daily mean values of CO_2 flux were obtained by averaging all successful results for each measurement day. The values range from 8.3 ± 2.1 (24 June) to 18.1 ± 4.5 (25 June) Kg/s and corresponded to daily emissions of 718 and 1565 tons, in that order. These data are in agreement with past carbon dioxide estimates carried out at Stromboli. Moreover, in Figure 4b, our lidar-based CO_2 fluxes were compared with alternative measurements, obtained by multiplying the CO_2/SO_2 ratio by the SO_2 flux, which is obtained using the traditional Multi-GAS + UV spectroscopy-based techniques. Despite the issues of the Multi-GAS + SO_2 flux approach reported in the literature (distinct temporal resolutions, and poor temporal alignment [41]), we found a general agreement between the CO_2 fluxes obtained by BILLI and the usual instruments. This provided reciprocal trust for both techniques. Furthermore, BILLI-based carbon dioxide fluxes showed a better temporal resolution (≈16–33 min) and greater continuity. Moreover, BILLI as well as other standoff instruments, is intrinsically safe for operators [41].

In conclusion, although several improvements were made (higher distances reached, fully automated operational routines, independent measurements of plume transport velocity) with respect to the previous campaigns [39–41], efforts were needed to improve the system (in particular, improving portability and reducing power requirements). At the same time, further in-field tests were still necessary to fully explore the potential of our system, validating the reliability and efficiency of the data analysis method.

3.3. The Experimental Campaign of Mt. Etna

The last experimental campaign was carried out at Mt. Etna Volcano (3329 m ASL), a hostile and only partially inhabited region located near Catania, in Sicily. Mt. Etna is the largest and most important volcano in Italy and one of the most active volcanoes in the world. For these reasons, it was the ideal place to test our system performances [42]. Similar to previous campaigns, the main goal was to measure the exceedance of in-plume CO_2 concentration and flux to provide useful information to volcanological research for a precocious alert to the population in the case of eruptions.

The area chosen for our experiments was the northeast summit crater of the volcano (other craters were discarded due to fact that volcanic plume is too dilute or only partially visible) and is characterized by extremes of temperature, high humidity, and rates of rainfall, and the presence of acid vapors, re-suspended dust and particles, which are toxic for humans and dangerous to the instrumentation. Notwithstanding the harsh environment, BILLI operated practically continuously for nearly a week (from 28 July to 1 August 2016 including an initial instrumental setup phase). During our experiment, the system was placed on a trailer, loaded on a truck, positioned into the courtyard of the INGV (Italian National Institute of Geophysics and Volcanology)observatory "Pizzi Deneri" at 2823 m ASL, on the northeast side of the main crater. This fixed position, about 500 m below the volcano summit, was ≈3 km far from the most important craters [42]. Changing the elevation angle in the range 7–14° while fixing the azimuth angle (230°), BILLI vertically scanned the plume. Atmospheric profiles were recorded every 10 s, whereas each scan, obtained combining 24 profiles, was completed in ~15 min.

Taking into account the local conditions, both the setup (e.g., ON/OFF wavelengths) of the system and the data processing routine remained the same as the previous campaign, with the only exception of a reduction in the system weight (thanks to a rearrangement of the mechanical frame of the system). For these reasons, in this case, laser beam also reached longer ranges and the systematic uncertainty remained stable at 5.5%, whereas, as reported in the literature [39], the statistical uncertainty of lidar profiles increased with range. At 2.5 km, a mean range, it was about 2% [41,42], while it exceeded 5% at 4.2 km [42]. Thanks to the measures already adopted in the previous campaign and reported in Section 2.1, the field setup at Mt. Etna, and the good weather conditions, a slight improvement in the system performances was observed.

Thanks to the BRIDGE DIAL technique, it was possible to retrieve up to 220 scans (lasting 10–20 min) of CO_2 concentration in excess in the plume and, at the same time, the CO_2 flux time-series. The whole dataset covered six days of operation, for a maximum of 12 consecutive hours. Examples of CO_2 profiles, dispersion maps, and flux time-series are reported in Figure 5a–c, in that order [40].

During the campaign, several configurations of plume were recognized, which varied as a function of acquisition time. In fact, both the magnitude and the size of the CO_2 plume were subject to change with the elevation of the laser beam pointing angle and, obviously, they were also influenced by the presence of wind [42].

In general, for elevation angle values comprised within 7° and 9°, it was quite common to detect narrow plume peaks (extending for about 500 m with magnitude values not exceeding 140 ppm) attached or strictly close to the first or the second rockface of Mt. Etna. In fact, at that elevation angle range, the laser beam altitude was lower than the summit crater. Therefore, the profiles' behavior could be justified by the proximity of the rockface behind the plume that, de facto, prevented the complete dispersion of the volcanic cloud into the atmosphere and, at the same time, made the wind contribution negligible.

The situation changed by increasing the elevation angle, as is shown by the profile reported in Figure 5a, which was acquired in the free atmosphere at 15:19 (local civil time) with a 12° elevation angle and a 230° azimuth angle [42]. In Figure 5a, it is possible to note two peaks, relative to CO_2 plume, highlighted by the red boxes. As expected from previous considerations, the first peak, localized between 1250 and 2550 m, was extremely wide (it extends for more than 1 km) and its magnitude exceeded the value of 150 ppm. Instead, the second plume peak, around 250 ppm, was detected beyond 4 km (precisely from 4060 m to 4140 m) from the system location and represents the farthest plume peak ever detected by our system. The magnifications of both plumes are clearly visible in the lower part of the figure. The measurement uncertainty (obtained by the quadratic sum of both statistical and systematic uncertainties) is equal to 6% and 7.5%, respectively [42]. Instead, the portion of the profile between the two mentioned plume peaks showed several oscillations. These latter were due to the residual presence of instrument noise. For this reason, it was neglected. The behavior of this plume could be justified by the fact that, for such elevation angles and altitudes, the plume was completely dispersed in the free atmosphere and scattered by a moderate presence of the wind [42].

Figure 5b shows an example of CO_2 dispersion vertical map, acquired on 31 July 2016, from 12:47 p.m. to 1:03 p.m. This time slot was selected for the relevance of the results. The tomography represents the in-plume concentration exceedance as a function of elevation angle and range of acquisition. The time evolution of the plume is clearly visible, highlighted by the red, orange, yellow, and green spots with respect to the dark blue color that, instead, represents the CO_2 concentration of the natural background. As already stressed, at lower elevation angles, the plume was circumscribed between the first and second rockface of Mt. Etna encountered by the laser beam. The rockface regions have been highlighted by the semitransparent yellow dotted boxes in the lower side of each scan. Instead, in the free atmosphere, the smoke plume rose and developed in both the vertical and horizontal direction, covering a wide area, after a complete random dispersion. Plume fluctuations in the free atmosphere were also probably due to a combination of the thermal updraft, from the main degassing vents on the volcano summit, and on a smaller scale, to the presence of northwest wind blowing during the measurement session. Furthermore, differences between subsequent maps were also probably due to rapid, local, and random fluctuations of particles and gases emitted by the volcano [42].

Dispersion maps such as the one shown in Figure 5b provide the basis for the computation of the volcanic carbon dioxide flux. In Figure 5c, an example of CO_2 flux series, lasting 6 h (from 12:22 p.m. to 6:08 p.m. local civil time), is reported. As in the Stromboli volcano campaign, CO_2 flux was obtained by multiplying the plume transport velocity with the plume CO_2 molecular density. The first term was obtained with a permanent UV camera network, while the second one was inferred by summing the carbon dioxide excess over the cross section of the volcanic plume [42]. Then, following the procedure reported in Section 2.2.4, it was possible to retrieve the CO_2 flux. The CO_2 flux varied from 13 to 92 kg/s (1235 to more than 8000 tons/day) during the measurement interval, with an average value of 40 ± 13 kg/s (3900 ± 1287 tons/day). The bars in Figure 5c show the CO_2 flux errors ($\approx$30%), and were calculated propagating the errors of plume velocity and CO_2 concentration (procedure detailed in [41,42]). The northeast crater exhibited quiescent degassing during the whole lidar operation. For this reason, the usual variations in degassing were regarded as a consequence of the fluctuations in the magma/gas transport rate occurring in the feeding conduits of Mt. Etna [42].

In order to add confidence to our results, the lidar-based CO_2 flux time-series were compared with independent measurements based on the usual techniques that involve determinations of SO_2 fluxes and CO_2/SO_2 ratios relative to the plume. The averaged values of CO_2 flux retrieved by conventional techniques was found to be equal to about 2750 ktons/day, in good agreement with our measurements. Furthermore, our CO_2 flux time-series (see Figure 5c) were found to be close to the historical time-series retrieved by independent in situ observations: at Mt. Etna, on average, more than 2 ktons/day of carbon dioxide are outputted every day in periods of quiescent passive degassing, while its carbon dioxide emission rates were from 10 to 40 times larger during the effusive event of 2004–2005 [42].

In conclusion, the great novelty, with respect to the previous works [21,37–41], was that our measurements allowed us to locate and track volcanic plumes beyond 4 km of distance from the system location with a lower uncertainty level with respect to the first field tests. In addition, our detected values in excess of CO_2 concentration were consistent with both the conventional measurements (according to volcanologists information) carried out in the same time interval, and with the previous ones acquired during the Stromboli campaign [39–42].

The results reported in this section can be considered to be extremely promising to validate the reliability and accuracy of our system. Furthermore, they represent a further step forward in ground-based volcano monitoring and volcanological research.

4. Conclusions

The main goal of this paper was to show the potential of mobile DIAL-Lidar systems for volcanological research such as the one reported in this work, designed for the remote measurement of both in-plume exceedance of volcanic CO_2 concentration and flux, considered as precursors of volcanic eruptions. Considering the difficulties and, in some circumstances, the hazardousness in retrieving these parameters as well as the limits of conventional in situ instruments, the BILLI system and the related BRIDGE DIAL data analysis technique have been developed. Here, particular emphasis has been placed on the data collection and processing techniques, and the most significant findings of the experimental campaigns carried out over three years at the most hazardous Italian volcanic areas.

Observing the results related to the three experimental campaigns, carried out from October 2014 to July 2016, it is possible to note a gradual refinement of the system performance and an improvement in the accuracy of results. In fact, the implementation of the photo-acoustic cell has made it possible to reduce the systematic uncertainty of CO_2 concentrations to 5.5% and to evaluate a cumulative CO_2 measurement uncertainty (considering both the statistical and systematic uncertainty) between 6% and 7.5%.

The reported approach for the analysis of acquired data allowed us to retrieve, accurately and rapidly, range-resolved CO_2 concentration profiles and maps. Furthermore, and with the aim to improve the already high potential of our system, a further data processing routine was implemented and tested for independent measurements of the plume transport velocity. Nevertheless, this technique, although promising, still requires efforts to be refined. In this way, BILLI will be able to perform simultaneously measurements of CO_2 concentration, CO_2 flux and wind velocity, without additional costs or the support of other conventional devices for the same purpose.

Both CO_2 concentration profiles and maps are fundamental to describe the spatio-temporal dispersion of the volcanic plume into the atmosphere. CO_2 flux time-series retrieved by BILLI (affected by an overall uncertainty of $\approx$25%) were extremely useful to volcanologists for a fast tracking of volcanic activity, particularly if compared with conventional acquisitions.

The detected values in excess of CO_2 concentrations and fluxes were in good agreement with conventional measurements, carried out in the same time interval, but based on completely independent and significantly different approaches. This has proven the sensitivity and reliability of our system for the detection and monitoring of CO_2, not only in limited areas, but also over extended regions and for prolonged periods of time.

In summary, the novelty/advantage of this work is that the proposed standoff DIAL system allows effective measurements (CO_2 excess of a few tens of ppm has been clearly detected) to be taken continuously and remotely (up to more than 4 km from the crater), therefore, from a safer location free from the risks to which operators are exposed during direct sampling. In addition, data have been acquired with much higher temporal (10 s) and spatial (5 m) resolution than conventional instruments (the plume was scanned in few minutes rather than over several hours). Then, a complete time-resolved plume evolution has been detected in several measurement sessions; this can be useful for the assessment of the wind velocity, a crucial parameter not only for the CO_2 flux retrieval, but also for air traffic control applications. These performances allow the operator to characterize the spatio-temporal evolution of the plume, thus providing—24 h and real time—accurate information on an important precursor of eruptions.

To our knowledge, our system performed, for the first time, range-resolved measurements of CO_2 concentration and flux in a volcanic plume and the findings reported laid down the basis for a new and smarter generation of active optical devices, specifically conceived for long-term volcanic gas observation. Even though several improvements have been made (high distances reached, fully automated operational routines, reduction of system weight, and independent measurements of wind velocity), further developments are still necessary before DIAL-Lidar systems can become operative tools for real-time volcano monitoring. Efforts will have to be made, in particular, to improve portability (the current system weight is about 1100 kg), to reduce power requirements (currently, 6.5 KW) and to integrate existing system control and data processing routines to create a unique, user-friendly, and fully automated software framework for the end-users.

In the near future, it will be desirable to deploy both ground-based (using a newly designed laser-based system such as the one described in this paper together with conventional sensors) and airborne (or space borne) platforms, forming a robust, integrated network for eruptions forecasting for prolonged periods. This will allow us to provide useful information to volcanologists, concerning the time evolution of volcanic gases in hazardous regions while working remotely and safely.

Author Contributions: L.F., S.S., and A.A. conceived and designed the experiments; S.S., L.F., M.N., and A.A performed the experiments; S.P. and G.M. analyzed the data; S.P., L.F., and S.S. wrote the paper with contributions from all co-authors. All authors have read and agreed to the published version of the manuscript.

Funding: The research leading to these results received funding from the European Research Council under the European Union's Seventh Framework Program (FP7/2007/2013)/ERC, grant agreement n 305377 (PI, Aiuppa).

Acknowledgments: The authors are grateful to ENEA, in general, and Aldo Pizzuto, Roberta Fantoni, and Antonio Palucci, in particular, for their constant encouragement. We also thank the staff of INGV-OE, and especially the Director Eugenio Privitera and Salvatore Consoli for the logistical support and granting access to the INGV observatory "Pizzi Deneri" at Mt. Etna. Società Enel Produzione spa is kindly acknowledged for logistical support and for providing access to "Centrale ENEL di Stromboli" during the field campaign. The authors wish to thank the Comune di Lipari, Circoscrizione di Stromboli, for authorizing the use of the lidar on Stromboli. The support from the ERC project BRIDGE, no. 305377, is gratefully acknowledged.

Conflicts of Interest: The authors declare no conflict of interest.

References

1. Paik, S.; Min, S.K. Assessing the Impact of Volcanic Eruptions on Climate Extremes Using CMIP5 Models. *J. Clim.* **2018**, *31*, 5333–5349. [CrossRef]
2. Sparks, R.S.J.; Biggs, J.; Neuberg, J.W. Monitoring Volcanoes. *Science* **2012**, *335*, 1310–1311. [CrossRef] [PubMed]
3. Chouet, B.; Matoza, R. A multi-decadal view of seismic methods for detecting precursors of magma movement and eruption. *J. Volcanol. Geotherm. Res.* **2013**, *252*, 108–175. [CrossRef]
4. Dzurisin, D. A comprehensive approach to monitoring volcano deformation as a window on the eruption cycle. *Rev. Geophys.* **2003**, *41*, 1001–1030. [CrossRef]
5. Biggs, J.; Ebmeier, S.K.; Aspinall, W.P.; Lu, Z.; Pritchard, M.E.; Sparks, R.S.J.; Mather, T.A. Global link between deformation and volcanic eruption quantified by satellite imagery. *Nat. Commun.* **2014**, *5*, 3471. [CrossRef]
6. Iverson, R.M.; Dzurisin, D.; Gardner, C.A.; Gerlach, T.M.; Lahusen, R.G.; Lisowski, M.; Major, J.; Malone, S.D.; Messerich, J.A.; Moran, S.C.; et al. Dynamics of seismogenic volcanic extrusion at Mount St Helens in 2004–05. *Nature* **2006**, *444*, 439–443. [CrossRef]
7. Aiuppa, A. Volcanic gas monitoring. In *Volcanism and Global Environmental Change*; Schmidt, A., Fristad, K.E., Elkins-Tanton, L.T., Eds.; Cambridge University Press: Cambridge, UK, 2015.
8. Oppenheimer, C.; Fischer, T.; Scaillet, B. Volcanic Degassing: Process and Impact. In *Treatise on Geochemistry*, 2nd ed.; Holland, H.D., Turekian, K.K., Eds.; Elsevier: Oxford, UK, 2014; Volume 4, pp. 111–179. [CrossRef]
9. Edmonds, M. New geochemical insights into volcanic degassing. *Philos. Trans. R. Soc. A* **2008**, *366*, 1885. [CrossRef]
10. Aiuppa, A.; Burton, M.; Caltabiano, T.; Giudice, G.; Guerrieri, S.; Liuzzo, M.; Murè, F.; Salerno, G. Unusually large magmatic CO_2 gas emissions prior to a basaltic paroxysm. *Geophys. Res. Lett.* **2010**, *37*. [CrossRef]
11. Burton, M.R.; Sawyer, G.M.; Granieri, D. Deep Carbon Emissions from Volcanoes. *Rev. Miner. Geochem.* **2013**, *75*, 323–354. [CrossRef]

12. Werner, C.; Fischer, T.P.; Aiuppa, A.; Edmonds, M.; Cardellini, C.; Carn, S.; Chiodini, G.; Cottrell, E.; Burton, M.; Shinohara, H.; et al. Carbon dioxide emissions from subaerial volcanic regions: Two decades in review. In *Deep Carbon Past to Present*; Orcutt, B.N., Daniel, I., Dasgupta, R., Eds.; Cambridge University Press: Cambridge, UK, 2019; Volume 8, pp. 188–236.

13. Fischer, T.P.; Arellano, S.; Carn, S.; Aiuppa, A.; Galle, B.; Allard, P.; Lopez, T.; Shinohara, H.; Kelly, P.; Werner, C.; et al. The emissions of CO_2 and other volatiles from the world's subaerial volcanoes. *Sci. Rep.* **2019**, *9*, 18716. [CrossRef]

14. Queißer, M.; Burton, M.; Kazahaya, R. Insights into geological processes with CO_2 remote sensing—A review of technology and applications. *Earth Sci. Rev.* **2019**, *188*, 389–426. [CrossRef]

15. Platt, U.; Bobrowski, N.; Butz, A. Ground-Based Remote Sensing and Imaging of Volcanic Gases and Quantitative Determination of Multi-Species Emission Fluxes. *Geosciences* **2018**, *8*, 44. [CrossRef]

16. Oppenheimer, C. Ultraviolet sensing of volcanic sulfur emissions. *Elements* **2010**, *6*, 87–92. [CrossRef]

17. Theys, N.; Hedelt, P.; De Smedt, I.; Lerot, C.; Yu, H.; Vlietinck, J.; Pedergnana, M.; Arellano, S.; Galle, B.; Fernandez, D.; et al. Global monitoring of volcanic SO_2 degassing with unprecedented resolution from TROPOMI onboard Sentinel-5 Precursor. *Sci. Rep.* **2019**, *9*, 2643. [CrossRef] [PubMed]

18. Carn, S.; Fioletov, V.E.; McLinden, C.A.; Li, C.; Krotkov, N.A. A decade of global volcanic SO_2 emissions measured from space. *Sci. Rep.* **2017**, *7*, 44095. [CrossRef] [PubMed]

19. Allard, P.; Burton, M.R.; Mure, F. Spectroscopic evidence for a lava fountain driven by previously accumulated magmatic gas. *Nature* **2005**, *433*, 407–410. [CrossRef]

20. Fischer, T.P.; Chiodini, G. Volcanic, Magmatic and Hydrothermal Gas Discharges. In *Encyclopedia of Volcanoes*, 2nd ed.; Sigurdsson, H., Houghton, B., McNutt, S., Rymer, H., Stix, J., Eds.; Elsevier: Amsterdam, The Netherlands, 2015; pp. 779–797. [CrossRef]

21. Aiuppa, A.; Fiorani, L.; Santoro, S.; Parracino, S.; Nuvoli, M.; Chiodini, G.; Minopoli, C.; Tamburello, G. New ground-based lidar enables volcanic CO_2 flux measurements. *Sci. Rep.* **2015**, *5*, 13614. [CrossRef]

22. Conde, V.; Robidoux, P.; Avard, G.; Galle, B.; Aiuppa, A.; Muñoz, A.; Giudice, G. Measurements of SO_2 and CO_2 by combining DOAS, Multi-GAS and FTIR: Study cases from Turrialba and Telica volcanoes. *Int. J. Earth Sci.* **2014**, *103*, 2335–2347. [CrossRef]

23. Saccorotti, G.; Iguchi, M.; Aiuppa, A. In situ Volcano Monitoring: Present and Future in Volcanic Hazards, Risks and Disasters. In *Volcanic Hazards, Risks, and Disasters*; Shroder, J.F., Papale, P., Eds.; Elsevier: Amsterdam, The Netherlands, 2014; Volume 202, p. 169. [CrossRef]

24. Symonds, R.B.; Rose, W.I.; Bluth, G.J.S.; Gerlach, T.M. Volcanic-gas studies: Methods, results and applications. *Rev. Mineral.* **1994**, *30*, 1–66.

25. Aiuppa, A.; Fischer, T.P.; Plank, T.; Bani, P. CO_2 flux emissions from the Earth's most actively degassing volcanoes, 2005–2015. *Sci. Rep.* **2019**, *9*, 5442. [CrossRef]

26. Aiuppa, A.; Giudice, G.; Gurrieri, S.; Liuzzo, M.; Burton, M.; Caltabiano, T.; Mcgonigle, A.J.S.; Salerno, G.; Shinohara, H.; Valenza, M. Total volatile flux from Mount Etna. *Geophys. Res. Lett.* **2008**, *35*, L24302. [CrossRef]

27. Casadevall, T.J.; Rose, W.I.; Fuller, W.H.; Hunt, W.H.; Hart, M.A.; Moyers, J.L.; Woods, D.C.; Chuan, R.L.; Friend, J.P. Sulfur Dioxide and Particles in Quiescent Volcanic Plumes from Poàs, Arenal, and Colima Volcanos, Costa Rica and Mexico. *J. Geophys. Res.* **1984**, *89*, 9633–9641. [CrossRef]

28. Edner, H.; Ragnarson, P.; Wallinder, E.; Ferrara, R.; Cioni, R.; Taddeucci, G.; Svanberg, S.; Raco, B. Total fluxes of sulfur dioxide from the Italian Volcanoes Etna, Stromboli and Vulcano measured by differential absorption lidar and passive differential optical absorption spectroscopy. *J. Geophys. Res.* **1994**, *99*, 18827–18838. [CrossRef]

29. Porter, J.N.; Horton, K.A.; Mouginis-Mark, P.J.; Lienert, B.; Sharma, S.K.; Lau, E.; Sutton, A.J.; Elias, T.; Oppenheimer, C. Sun photometer and lidar measurements of the plume from the Hawaii Kilauea Volcano Pu'u O'o vent: Aerosol flux and SO_2 lifetime. *Geophys. Res. Lett.* **2002**, *29*, 30–34. [CrossRef]

30. Hobbs, P.V.; Radke, L.F.; Lyons, J.H.; Ferek, R.J.; Coffman, D.J.; Casadevall, T.J. Airborne measurements of particle and gas emissions from the 1990 volcanic eruptions of Mount Redoubt. *J. Geophys. Res.* **1991**, *96*, 18735–18752. [CrossRef]

31. Fiorani, L.; Colao, F.; Palucci, A. Measurement of Mount Etna plume by CO_2-laser-based lidar. *Opt. Lett.* **2009**, *34*, 800–802. [CrossRef]

32. Jäger, H. The Pinatubo eruption cloud observed by lidar at Garmisch-Partenkirchen. *Geophys. Res. Lett.* **1992**, *19*, 191–194. [CrossRef]

33. Weibring, P.; Edner, H.; Svanberg, S.; Cecchi, G.; Pantani, L.; Ferrara, R.; Caltabiano, T. Monitoring of volcanic sulphur dioxide emissions using differential absorption lidar (DIAL), differential optical absorption spectroscopy (DOAS), and correlation spectroscopy (COSPEC). *Appl. Phys. B* **1998**, *67*, 419–426. [CrossRef]

34. Fiorani, L.; Colao, F.; Palucci, A.; Poreh, D.; Aiuppa, A.; Giudice, G. First-time lidar measurement of water vapor flux in a volcanic plume. *Opt. Commun.* **2011**, *284*, 1295–1298. [CrossRef]

35. Winker, D.M.; Liu, Z.; Omar, A.; Tackett, J.; Fairlie, D. CALIOP observations of the transport of ash from the Eyjafjallajökull volcano in April 2010. *J. Geophys. Res.* **2012**, *117*, D00U15. [CrossRef]

36. Fiorani, L.; Aiuppa, A.; Angelini, F.; Borelli, R.; Del Franco, M.; Murra, D.; Pistilli, M.; Puiu, A.; Santoro, S. Lidar sounding of volcanic plumes. In Proceedings of the SPIE 8894, Lidar Technologies, Techniques, and Measurements for Atmospheric Remote Sensing IX, Dresden, Germany, 22 October 2013; p. 889407. [CrossRef]

37. Fiorani, L.; Santoro, S.; Parracino, S.; Nuvoli, M.; Minopoli, C.; Aiuppa, A. Volcanic CO_2 detection with a DFM/OPA-based lidar. *Opt. Lett.* **2015**, *40*, 1034–1036. [CrossRef] [PubMed]

38. Fiorani, L.; Santoro, S.; Parracino, S.; Maio, G.; Del Franco, M.; Aiuppa, A. Lidar detection of carbon dioxide in volcanic plumes. In Proceedings of the SPIE 9535, Third International Conference on Remote Sensing and Geoinformation of the Environment (RSCy2015), 95350N, Paphos, Cyprus, 19 June 2015. [CrossRef]

39. Fiorani, L.; Santoro, S.; Parracino, S.; Maio, G.; Nuvoli, M.; Aiuppa, A. Early detection of volcanic hazard by lidar measurement of carbon dioxide. *Nat. Hazards* **2016**, *83*, S21–S29. [CrossRef]

40. Parracino, S.; Santoro, S.; Maio, G.; Nuvoli, M.; Aiuppa, A.; Fiorani, L. Fast tracking of wind velocity with a differential absorption LiDAR system: First results of an experimental campaign at Stromboli volcano. *Opt. Eng.* **2017**, *56*. [CrossRef]

41. Aiuppa, A.; Fiorani, L.; Santoro, S.; Parracino, S.; D'Aleo, R.; Liuzzo, M.; Maio, G.; Nuvoli, M. New Advances in Dial-Lidar-based remote sensing of the volcanic CO_2 flux. *Front. Earth Sci.* **2017**, *5*, 15. [CrossRef]

42. Santoro, S.; Parracino, S.; Fiorani, L.; D'Aleo, R.; Di Ferdinando, E.; Giudice, G.; Maio, G.; Nuvoli, M.; Aiuppa, A. Volcanic Plume CO_2 Flux Measurements at Mount Etna by Mobile Differential Absorption Lidar. *Geosciences* **2017**, *7*, 9. [CrossRef]

43. Fiorani, L. Lidar application to litosphere, hydrosphere and atmosphere. In *Progress in Laser and Electro-Optics Research*; Koslovskiy, V.V., Ed.; Nova: New York, NY, USA, 2010; pp. 21–75.

44. Kovalev, V.; Eichenger, W.E. *Elastic Lidar, Theory, Practice and Analysis Methods*; Wiley & Sons, Inc.: Hoboken, NJ, USA, 2004.

45. Gordon, I.E.; Rothman, L.S.; Hill, C.; Kochanov, R.; Tan, Y.; Bernath, P.; Birk, M.; Boudon, V.; Campargue, A.; Chance, K.; et al. The HITRAN 2016 molecular spectroscopic database. *J. Quant. Spectrosc. Radiat. Transf.* **2017**, *203*, 3–69. [CrossRef]

46. Schafer, R.W. What Is a Savitzky-Golay Filter? *IEEE Signal Process Mag.* **2011**, *28*, 111–117. [CrossRef]

47. Eloranta, E.W.; King, J.M.; Weinman, J.A. The determination of wind velocitys in the boundary layer by monostatic LiDAR. *J. Appl. Meteor.* **1975**, *14*, 1485–1489. [CrossRef]

48. Papoulis, A.; Pillai, U. *Probability, Random Variables and Stochastic Processes*, 4th ed.; McGraw-Hill: New York, NY, USA, 2002.

MDPI
St. Alban-Anlage 66
4052 Basel
Switzerland
Tel. +41 61 683 77 34
Fax +41 61 302 89 18
www.mdpi.com

Applied Sciences Editorial Office
E-mail: applsci@mdpi.com
www.mdpi.com/journal/applsci